Corporate Governance, Corporate Social Responsibility and Sustainability Practices in Oil Companies

Published by
ACADEMIC FOUNDATION
in association with

Institute of Public Enterprise

ABOUT THE AUTHORS

Geeta Rani Duppati is Senior Lecturer, Department of Finance, Waikato University, New Zealand and also Visiting Professor, Institute of Public Enterprise (IPE), Hyderabad, India. Apart from teaching, her research interests include: performance appraisal of state level public enterprises, corporate governance, public and private enterprises, mergers and acquisitions and CSR/ sustainability studies. She has taught at undergraduate, graduate and postgraduate levels at Addis Ababa University and Debub University in Ethiopia; Asmara University and Asmara Commercial College in Asmara, Eritrea, and Aukland University of Technology, Auckland and University of Waikato, New Zealand.

R.K. Mishra, Senior Professor and Director, IPE, Hyderabad, did his PhD from University of Rajasthan. A Visiting Fellow at London Business School (UK), he also studied for International Teachers programme at SDA Bocconi, Milan. Dr Mishra taught at the University of Bradford and was a Visiting Professor at Maison Des Sciences De L' Hommes, Paris and Faculty of Economics, University of Ljubljana, Slovenia. He is a member of the UN Task Force on International Standards of Excellence in Public Administration and Education. He is special invite on OECD Working Group on Privatization and Corporate Governance of SOEs and Asian Network of Corporate Governance. He is Non-Executive Director on some of the public and private enterprise boards in India. He has considerable experience of teaching and guiding research for doctoral degree in management, economics, public administration and commerce. His areas of interest include: corporate governance, international management, international finance, public-private partnership and restructuring and environmental administration.

Stuart Locke is Professor, Department of Finance, University of Waikato, New Zealand. Dr Locke has a background in finance, economics and chartered accountancy. His current research interests are in corporate governance in mature and emerging markets across the spectrum from small businesses, cooperatives, voluntary, state sector and listed public companies. He continues to publish widely in academic journals, participates regularly in media events, public speaking at conventions and is a Board member of his local Chamber of Commerce.

Corporate Governance, Corporate Social Responsibility and Sustainability Practices in Oil Companies

Evidence from Emerging Nations

GEETA RANI DUPPATI

R.K. MISHRA

STUART LOCKE

ACADEMIC FOUNDATION

NEW DELHI

www.academicfoundation.org

First published in 2015
by

ACADEMIC FOUNDATION
4772-73 / 23 Bharat Ram Road, (23 Ansari Road),
Darya Ganj, New Delhi - 110 002 (India).
Phones : 23245001 / 02 / 03 / 04.
Fax : +91-11-23245005.
E-mail : books@academicfoundation.com
www.academicfoundation.org

in association with
Institute of Public Enterprises (IPE), Hyderabad.

Disclaimer:
The findings/views/opinions expressed in this book are solely those of the authors and do not necessarily reflect the views of the publisher.

Cataloging in Publication Data--DK
Courtesy: D.K. Agencies (P) Ltd. <docinfo@dkagencies.com>

Geeta Rani, D., author.
Corporate governance, corporate social responsibility and sustainability practices in oil companies / Geeta Rani Duppati, R.K. Mishra, Stuart Locke.
pages cm
ISBN 9789332701656

1. Petroleum industry and trade--Management. 2. Petroleum industry and trade--Management--Case studies. 3. Corporate governance. 4. Social responsibility of business. 5. Sustainable development. I. Mishra, R. K., 1948-, author. II. Locke, Stuart M. (Stuart Murrin), 1953-, author. III. Institute of Public Enterprise (Hyderabad, India)issuing body. IV. Title.

DDC 338.27282068 23

Typeset by Italics India, New Delhi.

Printed and bound by The Book Mint, New Delhi.
www.thebookmint.in

Contents

List of Tables, Figures and Appendices

Tables

Figures

Appendices

Preface

In recent years, corporate governance has gained tremendous importance and interest among emerging and developing countries. This is due to the emergence of global norms for corporate governance and the developments brought about through globalisation, concerning harmonisation of procedures and structures. However, there is limited research done and literature on the topic of corporate governance in emerging economies compared to advanced economies. This is mostly due to the fact that developed countries have been the pioneers of the modern concept of corporate governance.

In 2013, for the first time since reliable records begin, developing and emerging countries are projected to have bigger economies than advanced and developed countries. Although this is only true for GDP (gross domestic produced) measured in PPP (purchasing power parity) terms, this will be a symbolic event that is expected to set the trend for decades to come. The world economy is expected to expand by 3.3 per cent in PPP terms and this time, compared to past cycles, emerging economies will be critical in driving the global economic growth. Research shows that India, China and Brazil are projected to contribute around half of the global economic growth in 2013 which contrasts with the 25 years preceding the financial crisis, when they contributed only around a quarter of global growth.

Due to the increased importance of emerging economies and the topic of corporate governance, this book, written from an academic perspective, aims to offer insight into the various issues surrounding corporate governance in the public and private sector in the emerging economies. It also touches on the issues of corporate fraud, forgery, debt, corruption and integrity in the world of corporate governance. The book is intended for practitioners as well as aspiring practitioners who are interested in gaining insight into the topic of corporate governance in the emerging economies.

1

Corporate Social Responsibility, Corporate Governance and Sustainability

Introduction

In the face of the current globalisation of economies and financial markets, and in an industrial climate characterised by restructuring, relocation and subcontracting, the social and ethical role of enterprises has come under increasing scrutiny. Multinational corporations (MNCs)[1] and local businesses are no longer able to conduct destructive and unethical practices, such as polluting the environment, without attracting negative feedback from the general public. With increased media attention, pressure from NGOs (non-governmental organisations) and rapid global information sharing, there is a surging demand from civil society, consumers, governments and others for corporations to conduct sustainable business practices.

This chapter presents an overview of corporate social responsibility (CSR), sustainability and corporate governance. Thereafter, relevance of these terms to the resource–based companies, i.e. oil and petroleum companies is reviewed and discussed.

Corporate Social Responsibility

Expectations of the internal and external players are growing on the business corporations. For example, employees want good

1 MNCs are those that conduct business in more than one country.

employers, customers want fair products, and society wants generous and environmentally sound companies. This indicates that businesses are no longer governed by shareholder value alone, but rather, their decisions are influenced by their responsibility towards employees, customers, the environment and the society they operate in. For a rising number of businesses, whether large or small, ignoring these 'soft' factors is weakening their competitive edge.

The driving force behind the rising significance of CSR is capital market expectations. Analysts and rating agencies no longer just look at revenue and profit, they are also starting to examine whether a company's profits were generated in a sustainable manner, that is: applying ecologically and socially sound principles. Employees, customers and the wider public is also starting to show an interest in how a company manages its business (Büchner, 2012).

In addition, in order to attract and retain employees and customers, many companies now realise the importance of being ethical while running their daily operations. The corporate response has often meant an adoption of 'a new consciousness,' and this has been known as CSR since the 1970s.

According to the Department of Public Enterprises (DPE), India, "Corporate Social Responsibility is a company's commitment to operate in an economically, socially and environmentally sustainable manner, while recognising the interests of its stakeholders" (DPE Guidelines on CSR for Central Public Sector Enterprises, 2010).

The American concept of CSR expresses the assumption that in the US business world, companies play a key role when it comes to social responsibility. Since only persons, rather than companies, can take on social responsibility, CSR presupposes a certain attitude on the part of a company's managers. In other words, CSR could also be described as a kind of value or management attitude that has sustainable business as its main objective and leads the company to engage in civic activities in its closer environment.

The idea of CSR initially emerged in North America during the 1960s and has only recently entered the spotlight in Europe. The CSR concept in Europe differs a great deal from the American understanding because many social issues that are part of the original CSR social approach, such as employee participation, education and health care, are regulated by law in European countries. Many elements of the CSR concept, formulated and theoretically developed in the US, were already implicitly part of European attitudes and laws long before the idea of explicitly formulated CSR strategies was brought to Europe (for the most part by MNCs). Therefore, the voluntary character of CSR and its recent rise to political and economic prominence is on the one hand seen critically—as a potential instrument for deregulation and undermining traditionally established labour regulation—but also as a chance to implement social practices which exceed what is required by existing laws. While in Germany, the industrial relations are significantly shaped by co-determination. Due to the strength of employee participation, they are more consensual than is the case in many countries, which benefits both working men and women, and their employers.

From an Indian perspective, CSR is viewed from narrow and also wider perspective. For instance, in the words of Rath (2012), society has emerged as a major stakeholder of the corporations. Business is seen as a social trust aimed at satisfying social expectations. The non-financial parameters like ethics, social responsibility and sustainable development are increasingly being regarded as factors responsible for lending a competitive edge to corporations. Social good in addition to profit is the basis for judging the performance of a business. CSR initiatives should be integrated into the business strategy of a corporation, and the culture in the organisation should be conducive to acceptance of social responsibility of the business.

According to Mishra *et al.* (2012), majority of the organisations in India are unable to envision the strategic return for enterprises that CSR can bring about in the form of long-term gains—generating more extensive return for responsible enterprises—and hence, do not link CSR with the strategy of the organisation.

From the context of Indian regulation on CSR, the recently enacted Companies Act (2013) that mandates the companies of a certain size spend two per cent of their three-year average annual profit towards CSR, is a landmark one as it makes India among the first nations to have social welfare spending as part of company statute by law (Dhawan, 2013). Around 8,000 companies would fall under the Bill's ambit and this mandate would translate into an estimated CSR spending of ₹12,000-15,000 crore annually. The Companies Act, 2013 states the following:

(1) Every company having net worth of ₹500 crore or more, or turnover of ₹1,000 crore or more or a net profit of ₹5 crore or more during any financial year shall constitute a CSR committee, with the board consisting of three or more directors, out of which at least one director shall be an independent director.

(2) The Board's report shall disclose the composition of the CSR committee.

(3) The CSR committee shall:

   (a) Formulate and recommend to the Board, a CSR policy which shall indicate the activities to be undertaken by the company.

   (b) Recommend the amount of expenditure to be incurred on the activities referred to in clause (a).

   (c) Monitor the CSR policy of the company from time to time.

(4) The Board of every company shall:

   (a) After taking into account the recommendations made by the CSR committee, approve the CSR policy for the company, disclose contents of the policy in its report, and also place it on the company's website, if any, in the manner as may be prescribed.

(b) Ensure that the activities as are included in CSR Policy of the company are undertaken by the company.

(5) The Board of every company shall ensure that the company spends, in every financial year, at least two per cent of the average net profits of the company made during the three immediately preceding financial years, in pursuance of its CSR policy: provided that the company shall give preference to the local area and areas around it where it operates, for spending the amount earmarked for CSR activities; provided further that if the company fails to spend such amount, the Board shall, in its report, specify the reasons for not spending the amount.

All over in western corporate sectors CSR has a new meaning, that of a contributor to the society rather than a mandatory activity. However, there are concerns that had been expressed about the CSR's spend being made mandatory. Prof Milton Friedman once opined that "the only social responsibility of business is to do business and increase the shareholders value." Companies now pay a superfluity of taxes without having the voting power. Industrial bigwigs argue that mandatory spending would encourage double taxation on shareholders. Management must not be forced to not spend so much as a single rupee on CSR. If the shareholders want CSR, they have to instruct the management to do so. Moreover, mandatory spending would definitely open new doors of unfair practices by corporate to hide profits. These are the opinions of the alarmed corporate world. Should the companies follow CSR and spend compulsorily is a policy decision that is to be determined by the government. These decisions should be made keeping national interests, social welfare and above all protection of stakeholders' interests as paramount.

Corporate Governance

Corporate governance can be defined as the study of power and influence over decision making within the corporation (Aguilera and Jackson, 2010). As per the existing literature, scholars have approached the subject of corporate governance from a variety of disciplinary perspectives, including economics, management, law, political science, culture and sociology. Aguilera and Jackson (2010) had reviewed a range of definitions on corporate governance, and had shown how the international comparison of corporate governance raises issues of comparing apples and oranges.

Existing definitions of corporate governance are closely tied to different paradigms or ways of conceptualising the organisation or firm. Economists have tended to see corporate governance as a nexus of contracts among owners, who essentially pursue private means and run the corporation in their own self-interest without explicit obligations to society. Thus, Zingales (1998) conceptualised corporate governance from a Williamsonian (1985: 496, italic in original) perspective as, "the set of constraints that shape the *ex post* bargaining over the quasi-rents generated in the course of a relationship." The key issue here is that contracts are incomplete, and there a lot of room for bargaining about how to divide the firm surpluses.

Other economists such as Shleifer and Vishny (1997) approached corporate governance from an agency perspective, arguing that corporate governance "deals with the ways in which suppliers of finance to corporations assure themselves of getting a return on their investment." This definition is very much tied to the idea of how shareholders in a managerial-controlled public firm will minimise their agency costs.

Agency problems have been central to research among management scholars. For example, Walsh and Seward (1990) compared internal versus external mechanisms of corporate control. More recently, an article by Dalton et al. (2007) on the fundamental agency problem offered insightful arguments on how the agency theory perspective affects different corporate governance mechanisms such as the board of directors, ownership structures and the market for corporate control. From a finance and economics perspective, and still with the agency theory as the driving force, Adams *et al.* (2010) reviewed the role of boards and the challenges of their comprehensive study as a central piece of the corporate governance policy debate.

Other approaches to corporate governance in economics and management are inspired by theories of the firm based on the unique nature of the employment relationship (Gospel and Pendleton, 2005). One alternative to the principle-agent view is the 'team production' model, which suggests that the corporation embodies a number of stakeholders who invest in firm-specific resources, but for their own benefit jointly relinquish control over those resources to a board of directors in order to solve the problem of coordinating efforts within the team (Blair and Stout, 1999). Along similar lines, the concept of essentiality has been used to elaborate the idea that where human assets are essential to the productivity of the firm, control based on ownership of the physical assets or legal entity of the corporation cannot act as a substitute for cooperation or employee voice in decisions (Aoki and Jackson, 2008).

Others have developed a wider theory of innovative enterprise, stressing how corporate governance may or may not support the strategic, organisational and financial perquisites of innovation (Lazonick, 2007; O'Sullivan, 2000). Again, these approaches define and explore corporate governance in relation to economic problems of making investments in the firm, but extend the view of organisations beyond agency theory to look at investments of human capital in particular. Finally, in management scholarship, stakeholder theory has become a useful framework to explain the complex and wider relationships among the different stakeholders in the firm (Donaldson and Preston, 1995; Freeman, 1984; Freeman *et al.*, 2010; Jones, 1995; Schneper and Guillen, 2004), a point that is also discussed in the CSR literature (Aguilera *et al.*, 2007; Margolis and Walsh, 2003).[2]

Sustainability and Sustainable Performance

According to Thomas and Nowak (2006), unlike CSR, sustainability refers not only to a company's responsibility towards its stakeholders, taking into account economic, ecological and social aspects, but also to its responsibility towards humanity as a whole, towards the environment and towards future generations. CSR is an integral element of sustainable corporate governance. Sustainable development envisions a development which takes economic, social and ecological aspects into account and treats them equally. Economic operations following this principle shall meet the needs of the present without compromising the ability of future generations to meet their own needs and the justified claims of the poorer countries to have access to wealth.

Corporate sustainability, is defined as meeting the needs of the present without compromising the ability of future generations to meet their own needs. Companies shaping corporate strategies with mutuality between needs of corporation and society practice corporate sustainability. Corporate sustainability is interpreted as meeting the needs of the present without compromising the ability of future generations to meet their own needs. Companies are shaping corporate strategies with mutuality between needs of corporation and society. Companies are able to achieve their business goals and increase long-term shareholder value by integrating economic, environmental and social opportunities into their business strategies.

2 Theories of CSR also argue that corporate decisions may have negative externalities on stakeholders who are not part of the decision-making. Thus CSR theories call for greater dialogue, participation, and responsibility toward stakeholders in company decision making (Vogel, 2006).

In 1987, the World Commission on Environment and Development sought to address the problem of conflicts between environment and development goals by formulating a definition of sustainable development, "Sustainable development is development which meets the needs of the present without compromising the ability of future generations to meet their own needs."

Three essential elements have been identified following the widespread discussion and usage of the concept of sustainable development (see, e.g., Holmberg, 1992; Reed, 1997) which includes:

- *Economic*: An economically sustainable system must be able to produce goods and services on a continuing basis, to maintain manageable levels of government and external debt, and to avoid extreme sectorial imbalances which damage agricultural or industrial production.
- *Environmental*: An environmentally sustainable system must maintain a stable resource base, avoid over-exploitation of renewable resource systems or environmental sink functions, and deplete non-renewable resources only to the extent that investment is made in adequate substitutes. This includes maintenance of biodiversity, atmospheric stability and other ecosystem functions not ordinarily classed as economic resources.
- *Social:* A socially sustainable system must achieve fairness in distribution and opportunity, adequate provision of social services including health and education, gender equity and political accountability and participation.

These three elements of sustainability introduce many potential complications to the original, simple definition of economic development. The goals expressed or implied are multidimensional, raising the issue of how to balance objectives and how to judge success or failure. For example, what if provision of adequate food and water supplies appears to require changes in land use that will decrease biodiversity? What if non-polluting energy sources are more expensive, thus increasing the burden on the poor, for whom they represent a larger proportion of daily expenditure? Which goal will take precedence?

Taken together, these principles clearly suggest new guidelines for the development process. They also require a modification of the goals of economic growth (Arrow *et al.*, 1995). The principles of sustainability imply new goals and policies in all major areas of economic development, one of which is Energy.

Energy: Both supply limits and environmental impacts, in particular of the accumulation of greenhouse gases, mean that it will be necessary to accomplish a transition away from fossil fuels well before 2050 (MacKenzie, 1996; IPCC, 2001). A non-fossil energy system would be significantly more decentralised, adapted to local conditions and taking advantage of opportunities for wind, biomass, and off-grid solar power systems. This is unlikely to occur without a major mobilisation of capital resources for renewable energy development in countries now rapidly expanding their energy systems (Johansson and José, 2002).

One is globalisation with a human face, as UN Secretary-General Kofi Annan has described the issue. Opening up world markets has, without doubt, created unprecedented global economic growth —with the promise of even more prosperity. Unfortunately, the benefits of this have not been shared equally across the globe. This is leading, understandably, to some questioning about the role and function of free markets.

Clearly, business believes in free and open markets, but markets must work for everyone. That is why the World Business Council for Sustainable Development (WBCSD) has accepted Kofi Annan's invitation to engage in the process of developing a 'Global Compact' between the UN and business to promote a shared set of core values.

The aim is to underpin globalisation and free, open markets with stable societies and a fair distribution of the benefits.[3]

Issues of Sustainability for Oil Companies

Oil companies face sustainability issues in each of the main areas mentioned. These issues can also be divided as environmental, development and governance issues (Frynas, 2009). This distinction is similar to the general division in economic, environmental and social issues mentioned before. Yet, it also entails political issues. Firstly, environmental issues arise due to the nature of operations in the oil industry. It can cause environmental damage in various ways, such as water and soil pollution, disturbance of community's flora and fauna, greenhouse effects, and waste problems. This is mainly caused by upstream activities, such as drilling in areas with high biodiversity, or by downstream activities, such as emissions into air, water or ground (Frynas, 2009).

Secondly, the development challenge mentioned by Frynas (2009) focuses mainly on international development and the ability of the private sector to reduce poverty and improve education by following socially responsible practices. Development issues are part of social issues in general faced by oil companies. Other social issues can be the negative effects of operations on local communities in developing countries, human rights, and health and safety issues (Jenkins, 2005).

This is also called resource dependency or the resource 'curse'. Corruption levels are high in resource-rich countries and political stability is negatively related with resource dependency (Frynas, 2009). The exports of natural resources could undermine good governance and political accountability in developing countries. The main issues affecting the petroleum refining industry are often seen as separate and contradictory. Economic, environmental, social and

3 "Sustainable Development and Business: As Governments Retreat from their Wide-ranging Role, the Private Sector must Step into the Breach", Björn Stigson, President of the WBCSD.

political issues are interconnected and complex. It should also be noted that economic sustainability entails long-term profitability and existence of the firm. The general aspects of the organisation need to be respected next to the sustainability issues mentioned (Baumgartner and Ebner, 2010).

Successful Corporate Cases with Sustainability Practices

On the brighter side, the successful businesses had been immune to the global cyclical changes and downturns. According to a survey conducted by David *et al.* (2013), 50 per cent of the survey respondents who have changed three or four business model elements say they profit from their sustainability activities, compared to only 37 per cent of those who changed only one element of their business model. 61 per cent of companies that have changed their business model and have sustainability as a permanent fixture on their management agenda says they have added profit from sustainability. More than 60 per cent of companies that have changed their business model and have sustainability as a permanent fixture on their management agenda say they have added profit from sustainability and found that top management attention is central. The discussion follows.

Finance: An Inclusive Model—A Paradigm Shift

According to Roche (2010), the experience of Hewlett Packard's CFO indicates challenges during the financial crisis in pulling together a big picture of all moving parts of the company, putting in place additional measures and getting everyone focused on tough environment. She says that there was a real demand placed on the finance function throughout 2009. The challenge wasn't just the recessionary environment; it was also the currency volatility. The late November–December 2008 period was very challenging because we'd get new forecasts showing massive moves in revenue, and obviously therefore in profit. Even through mid-2009, there were still some pretty big month-to-month jumps from a forecasting perspective. Revising the annual plan multiple times to get a sense of what was happening from a currency perspective wasn't something we could put the organisation through as often as we wanted, so we spent a lot of time coming up with new models to understand how the different businesses within HP would respond under different sets of circumstances.

We were in a position to help senior management really understand the dynamics that were going on—which gave finance a bigger voice at the table. It was a great learning experience for the business folks as well, because the finance people couldn't do it by themselves. They had to go and talk to people, and by asking the types of questions that the finance function asks, they got the business guys thinking as well. So it became a much more collaborative effort to deliver the new models and the new understanding of how businesses would respond under varied economic conditions.

Dell Inc: Strategising the situation by providing services is a case point. For instance, with the huge need for in-depth product profitability analysis, Dell has developed a purpose-built solution for manufacturers. The Dell Product & Profit Analytics (DPPA) solution allows business and IT leaders to make smarter business recommendations on their organisations' product mix based on fact-based profitability metrics. The solution is a mix of manufacturing-centric software, designed to connect with the real-world processes in manufacturing companies supply chain, inventory management, materials purchasing, transportation, manufacturing overhead and many others.

Timeliness of Dell Inc in developing purpose-built solutions for manufacturers is worth considering. To have a competitive advantage, Dell Inc is bringing an awareness of profitability in organisations. Making smart business decisions is a lot easier when manufacturers can uncover the truth about profitability for individual products, and even specific customers. New tools are emerging to help companies discover the true profitability picture, and make strategic decisions faster and with greater confidence.

The rules of the game are changing and the players have to re-align in their positions.

CSR Reporting: A Collaborative Model

Costco Retailer: According to Stone (2013), Costco Wholesale (Cost), the second-largest retailer in the U.S. behind Walmart, is an anomaly in an age marked by turmoil and downsizing. Known for its $55-a-year membership fee and its massive, austere warehouses stocked floor to ceiling with indulgent portions of everything from tilapia to toilet paper, Costco has thrived over the last five years. While competitors lost customers to the Internet and weathered a wave of investor pessimism, Costco's sales have grown 39 per cent and its stock price has doubled since 2009. The hot streak continued through last year's retirement of widely admired co-founder and Chief Executive Officer Jim Sinegal. The share price is up 30 per cent under the leadership of its new, plain-spoken CEO (Chief Executive Officer) Craig Jelinek.

According to Costco CEO Craig Jelinek, despite the sagging economy and challenges to the industry, Costco pays its hourly workers an average of $20.89 an hour, not including overtime (vs. the minimum wage of $7.25 an hour). By comparison, while Walmart's average wage for full-time employees in the US is $12.67 an hour, according to a letter it sent in April to activist Ralph Nader. Eighty-eight per cent of Costco employees have company-sponsored health insurance, while Walmart has 'more than half'. Costco workers with coverage pay premiums that amount to less than 10 per cent of the overall cost of their plans. It treats its employees well in the belief that a happier work environment will result in a more profitable company. "I just think people need to make a living wage with health benefits," says Jelinek. "It also puts more money back into the economy and creates a healthier country. It's really that simple."

Sustainability Practices: An Inclusive Model

According to Lamarre and Pergler (2009), risk-assessment processes typically expose only the most direct threats facing a company and neglect indirect ones that can have an equal or greater impact. For instance, a thunderstorm in a distant place wouldn't seem like cause for alarm. Yet in 2000, when a lightning strike from such a storm set off a fire at a microchip plant in New Mexico, it damaged millions of chips slated for use in mobile phones from a number of manufacturers. Some of them quickly shifted their sourcing to different US and Japanese suppliers, but others couldn't and lost hundreds of millions of dollars in sales. More recently, though few companies felt threatened by the outbreak of SARS (severe acute respiratory syndrome), its combined effects are reported to have decreased the GDPs (gross domestic products) of East Asian nations by two per cent in the second quarter of 2003.

And in early 2009, the expansion of a European public-transport system temporarily grounded to a halt when crucial component providers faced unexpected difficulties as a result of credit exposure to ailing North American automotive OEMs (original equipment manufactures). It is found that executives who systematically examine the way risks propagate across the whole value chain, including competitors, suppliers, distribution channels and customers, can foresee and prepare for second–order effects more successfully.

Auckland Airport: According to the Chairman Report (2013), the business strategy of Auckland Airport, 2009–2013 (which was Flight Path for Growth) has been a success. It significantly changed the business philosophy and performance and developed focus on 'making journeys better' for all customers and partners of Auckland Airport. The new five-year business strategy—faster, higher, stronger—responds to the challenges that were identified in changing aviation markets, changing customer expectations on top of competitive pressure on retail and a subdued commercial property market. According to Fryer (2013), Auckland Airport took a comprehensive if somewhat academic approach to implementing its first sustainability policy and plan. The approach proved incredibly effective at a time when knowledge and understanding amongst stakeholders was far more limited than it is today. It was important therefore, for the company to provide leadership in the sustainability space. Sound governance structures have supported engagement across the entire business. The result is a successful first five years of the business with 73 per cent of sustainability targets being achieved. It is interesting to note both the sustainability and environmental policies have been redrafted and together cover stakeholder expectations for environmental performance. The outcomes included: increase in productivity, processes, operations and performance.

Nestlé: Nestlé developed a cost model innovation at a novel point in its value chain. The company realised how to use coffee grounds— a by-product of its manufacturing process—to help power its factories. Instead of treating coffee grounds as waste, Nestlé discovered that burning them can generate steam, which is used in many of its factories. Some 60 per cent of the steam the company uses now comes from burning coffee grounds, significantly reducing Nestlé's reliance on natural gas. In addition, in over 20 years, the company was able to divert 1.24 million tons of coffee grounds away from landfills.

Role of Boards

The role of boards in failing organisations is a point of focus. The board of directors has a fiduciary duty to shareholders. They fulfil this duty by overseeing and monitoring the CEO's and the management's actions, vetoing poor decisions to protect the interest of the shareholders, and assisting and advising management in their effort to increase shareholders' wealth. The first role is notably important in failing organisations. The board of directors is perhaps the last line of defence against management misbehaviour and/or poor management decisions. Given that a large proportion of full blown crises and failures do not result from a single 'big bad' decision, but from an unbroken chain of bad decisions over a period of time, board of directors is supposed to draw upon their collective wisdom to spot some of the bad decisions and break the chain of errors before they spiral out of control. The second duty of the board of directors is an advisory one. Board members are expected to bring in knowledge and experience from their past managerial experiences and membership on other boards. The diverse experiences of the board of directors, it is hoped, would help solve challenging problems firms face during failure.

It is reasonable to argue that role of future boards of directors ought to be more vigilant and their contribution to the management of the firm needs to be increased. When facing a crisis or failure, some managers 'go for broke' by taking high risk decisions, while others act conservatively and 'stick to their knitting.' First, according to Kahneman and Tversky's (1979) prospect theory, managers become risk seeking in response to threat of failure. According to this theory, managers in failing firms behave like losing gamblers who do their wildest betting when they realise that there is a real chance of going broke (Mellahi, 2005). The more they lose, the wilder they bet. This is because when failure is perceived to be a real possibility, managers have little to lose and may take high-risk decisions that make matters worse. Similarly, a significant body of research shows that managers are likely to reinvest more in a project when they realise that the project is doing poorly. This is because managers are often unwilling to take a loss and as a result they often escalate their commitment to a failing course of action (Mellahi, 2005).

Staw noted that "because it is often possible for persons who have suffered a setback to recoup their losses through an even greater commitment of resources to this same course of action, a cycle of escalating commitment can be produced." As a result, when managers receive information that a project they champion is failing, they allocate more, rather than less, to the project. Rushed high-risk decisions, or throwing good money after bad because of escalation of commitment, are likely to make a bad situation worse. The role of the board of directors is to assess the risk of each strategy to stop management from gambling away the future of the organisation.

Second, according to threat-rigidity effect theory, some managers in threatening situations, such as a crisis or failure, tend to behave rigidly. Rather than conduct a serious analysis of the causes of failure and develop innovative strategies to deal with it, managers tend to cling to old ways of doing things and increase their use of cognitive simplification processes such as heuristics or 'rules of thumb.' The role of the board of directors is to challenge these maladaptive behaviours and advise management to focus on the real causes of failure and generate innovative strategies to turn the organisation around (Mellahi, 2005).

It is evident from the above that the proactive role of the boards with a positive approach towards sustainability practices will enable the corporate to sustain the crisis momentum. The following chapters will present and discuss the theories and models of corporate governance, considers five cases of oil companies from emerging nations and reviews their approaches and performance from the context of corporate governance practices, CSR approach and sustainability practices.

References

Adams, R.B., B.E. Hermalin and M.S. Weisbach (2010). "The Role of Boards of Directors in Corporate Governance: A Conceptual Framework and Survey", *Journal of Economic Literature* 48(1): 58-107.

Aguilera, R.V. and G. Jackson (2010). "Comparative and International Corporate Governance", *The Academy of Management Annals* 4(1), 485-556.

Aguilera, R.V., D.E. Rupp, C.A. Williams and I. Ganapathi (2007). "Putting the S Back in Corporate Social Responsibility: A Multilevel Theory of Social Change in Organizations", *Academy of Management Review* 32(3) 836-63.

Aoki, M. and G. Jackson (2008). "Understanding an Emergent Diversity of Corporate Governance and Organizational Srchitecture: An Essentiality-based Analysis", *Industrial and Corporate Change* 17(1), 1-27.

Arrow, K., B. Bolin, R. Costanza, P. Dasgupta, C. Folke, C.S. Holling, B.–O. Jansson, S. Levin, K.–G. Maler, C. Perrings and D. Pinentel (1995). "Economic Growth, Carrying Capacity, and the Environment", *Science* 268: 520-1.

Baumgartner, R.J. and D. Ebner (2010). "Corporate Sustainability Strategies: Sustainability Profiles and Maturity Levels", *Sustainable Development* 18: 76-89.

Blair, M.M. and L.A. Stout (1999). "A Team Production Theory of Corporate Law", *Virginia Law Review* 85(2): 247-328.

Büchner, L.M. (2012). "Corporate Social Responsibility and Sustainability from a Global, European and Corporate Perspective", *Eurolimes* 13 : 41-55.

Cameron, K.S., R.I. Sutton and D.A. Whetten (eds.) (1998). *Readings in Organizational Decline: Frameworks Research, and Prescriptions*. Cambridge, MA: Ballinger.

Collins, E., J. Roper and S. Lawrence (2010). "Sustainability Practices: Trends in New Zealand Businesses", *Business Strategy and the Environment* 19(8): 479-94.

Dalton, D.R., M.A. Hitt, Certo, S.T. and Dalton, C.M. (2007). "The Fundamental Agency Problem and its Mitigation: Independence, Equity, and the Market for Corporate Control", *Academy of Management Annals* 1: 1-64.

David, Nina, Reeves Kunt and Eugene (2013). "The Benefits of Sustainability-Driven Innovation", *MIT Sloan Management Review* 54(2): 69-73 (Winter).

Dhawan, Ashish (2013). "Companies Bill 2012: How Can India Inc Make CSR a Game-changer?", *Economic Times*, September 5.

Donaldson, T. and L. Preston (1995). "The Stakeholder Theory of the Corporation: Concepts, Evidence, and Implications", *Academy of Management Review* 20: 65-91.

DPE (2006). *Report of 2nd Pay Revision Committee for Executives of CPSEs*. New Delhi.

Fox, J. (2013). "How Shareholders are Ruining American Business", *The Atlantic*, 19th June.

Frank, M.Z. (1988). "An Intertemporal Model of Industrial Exit", *Quarterly Journal of Economics* 103: 333-44.

Freeman, R.E. (1984). *Strategic Management: A Stakeholder Approach*. Boston: Pitman.

Freeman, R.E., J.S. Harrison, A.C. Wicks, B.L. Parmar and S. de Colle (2010). *Stakeholder Theory: The State of the Art*. Cambridge: Cambridge University Press.

Fryer, M. (2013). *Case of Auckland Airport: Implementing Sustainable Business Practice at Auckland Airport—Drivers, Responses and Outcomes*. Massey University College of Business Sustainability Conference. 13-15 November. Albany Campus, Auckland, New Zealand.

Frynas, J. G. (2009). *Beyond Corporate Social Responsibility*. New York:Cambridge University Press.

Gail, Thomas and Margaret Nowak (2006). "Corporate Social Responsibility: A Definition", *Working Paper Series* No. 62. Curtin University of Technology, Graduate School of Business. December. *http://www.business.curtin.edu.au/files/GSB_Working_Paper_No._62_Corp_Social_Resp_A_def*

Gospel, H.F. and A. Pendleton (2005). *Corporate Governance and Labour Management: An International Comparison*. Oxford: Oxford University Press.

Harris, M.J., W. A. Timothy, G.P. Kevin and G.R. Neva (eds.) (2001). *A Survey of Sustainable Development: Social and Economic Dimensions* Washington, DC: Island Press.

Holmberg, J. (ed.) (1992). *Making Development Sustainable: Redefining Institutions, Policy, and Economics*. Washington, DC: Island Press

Intergovernmental Panel on Climate Change (IPCC). (2001). *Climate Change 2001: The Scientific Basis*. New York: Cambridge University Press.

Jean, P.R. (2013). "Europe's Crisis Will Hurt New Zealand", *Dominion Post Newspaper.*

Jenkins, R. (2005). "Globalization, Corporate Social Responsibility and Poverty", *International Affairs* 81: 525-40.

Johansson, T.B. and G. José (eds.) (2002). *Energy for Sustainable Development: A Policy Agenda*. New York: United Nations Development Programme.

Jones, T.M. (1995). "Instrumental Stakeholder Theory: A Synthesis of Ethics and Eonomics", *Academy of Management Review* 20: 305-60.

Jovanovic, B. and S. Lach (1989). "Entry, Exit and diffusion with Learning by Doing", *American Economic Review* 79: 690-99.

Kahneman, D. and A. Tversky (1979). "Prospect Theory: An Analysis of Decision under Risk", *Econometrica* 47(2).

Knott, A.M. and H.E. Posen (2005). "Is Failure Good?", *Strategic Management Journal* 26(7): 617-41 (Summer Special Issue).

Koller, T.M. (2010). "Why Value Value?", *McKinsey on Finance* 35: 2-8 (Spring).

Lamarre, E. and M. Pergler (2009). "Risk: Seeing Around the Corners", *McKinsey on Finance* 33.

Lawrence, S., E. Collins, K. Pavlovich and A. Murugesh (2006). "Sustainability Practices of SMEs: The Case of NZ", *Business Strategy and the Environment* 15(4): 242-57.

Lazonick, W. (2007). "The U.S. Stock Market and the Governance of Innovative Enterprise", *Industrial and Corporate Change* 16(6): 983-1035.

Levine, C.H. (1978). "Organizational Decline and Cutback Management", *Public Administration Review* 38: 316-25.

Lippman, S.A. and R.P. Rumelt (1982). "Uncertain Imitability: An Analysis of Inter Firm Differences in Efficiency under Competition", *The Bell Journal of Economics* 13: 418-38.

Locke, Stuart and Geeta Duppati (2013). "The Risk Adjusted Return on Indian Central Public Sector Enterprises", *Indian Journal of Corporate Governance*. Hyderabad: Institute of Public Enterprises.

Lozano, R. and D. Huisingh (2011). "Inter-linking Issues and Dimension in Sustainability Reporting", *Journal of Cleaner Production* 19: 99-107.

MacKenzie, J. (1996). *Oil as a Finite Resource: When will Global Production Peak?* Washington, DC: World Resources Institute.

Margolis, J.D. and J.P. Walsh (2003). "Misery Loves Companies: Rethinking Social Initiatives by Business", *Administrative Science Quarterly* 48(2): 268-305.

Mellahi, K. (2005). "The Dynamics of Boards of Directors in Failing Organizations", *Long Range Planning* 38(3): 261-79.

Mishra, R., S. Sarkar and P. Singh (2012). "Linking CSR to Business Strategy: The Balanced Scorecard Approach", Paper presented at the *Strategic Corporate Social Responsibility: Towards a Sustainable Business*. Hyderabad.

Ormerod, P. (2005). "Organizational Failure", *Long Range Planning* 38: 235.

O'Sullivan, M. (2000). "Corporate Governance and Globalization", *Annals of the American Academy of Political and Social Science* 570: 153-72.

Rath, A. (2012). "Integrating Corporate Social Responsibility with Business Strategy for Competitive Advantage", Paper presented at the *Strategic Corporate Social Responsibility: Towards a Sustainable Business*. Hyderabad.

Reed, D. (ed.) (1997). *Structural Adjustment, the Environment and Sustainable Development*. London: Earthscan Publications.

Schneper, W.D. and M.F. Guillen (2004). "Stakeholders Rights and Corporate Governance: A Cross-national Study of Hostile Takeovers", *Administrative Science Quarterly* 49(2): 263-95.

Shah, A. (2008). *Global Financial Crisis 2008*. Retireved from *http://www.globalissues.org/article/768/global-financial-crisis*

Sheth, J. and R. Sisodia (2005). "Why Good Companies Fail", *European Business Forum* 22: 25-30 (Autumn).

Shleifer, A. and R.W. Vishny (1997). "A Survey of Corporate Governance", *Journal of Finance* 52(2): 737-83.

Stone, Brad (2013). "Costco CEO Craig Jelinek Leads the Cheapest, Happiest Company in the World", *Bloomberg Businessweek,* 06 June.

Vogel, David (2006). *The Market for Virtue*. Washington, DC: Brookings Institution Press.

Walsh, J.P. and J.K. Seward (1990). "On the Efficiency of Internal and External Corporate Control Mechanisms", *Academy of Management Review* 15: 421-58.

Williamson, O.E. (1985). *The Economic Institutions of Capitalism*. New York: Free Press.

Witteloostuijn, A.V. (1998). "Bridging Behavioral and Economic Theories of Decline: Organizational Inertia, Strategic Competition, and Chronic Failure", *Management Science* 44: 501-19.

Yin, R.K. (2013). *Case Study Research: Designs and Methods*. Fifth Edition. Sage.

Zingales, L. (1998). "Corporate Governance", *The New Palgrave Dictionary of Economics and the Law*. London: Macmillan. pp.497-503.

CASE POINT

This section reviews two case studies:

- Auckland Airport and
- CSR Policies and Practices: A Case of India and New Zealand.

Case of Auckland Airport

This case presents the success story of Auckland Airport's approach to climate change, as well as its first five-year sustainability plan. In 2013, Auckland Airport Chair Joan Withers, and by extension the company itself, have been recognised for the leadership in the area of sustainability at the inaugural Asia-Pacific Economic Leadership Awards held in Singapore.

The awards recognise and honour the achievements of 'enterprise leaders' who are 'doing good' in the areas of sustainable development, corporate responsibility and CSR. Organised by the non-profit organisation Asia-Pacific Enterprise Cooperation Ltd (AP-EC), the awards were open to 16 Asia-Pacific territories, and the winners were announced at a black tie event at Marina Bay Sands in Singapore on 15 March. Auckland Airport sustainability and environmental manager Martin Fryer accepted the award on Joan's behalf and says it was an amazing experience to share the stage with some truly global companies such as Coca Cola and Procter and Gamble.

This study reviews the approach to embedding sustainable business practice at Auckland Airport, New Zealand's key international gateway and a transport hub that handles 14 million air passengers each year. It reflects on the successes and failures of the company's first sustainability policy and plan that covered the period 2008–2012. It also discusses the process improvements employed to create a second policy and plan setting the company targets out to 2020. The study looks at the outcomes of over six years concerted effort to operate and develop an airport in a sustainable manner and how these outcomes have been used to improve sustainability reporting and communicating performance to stakeholders. It begins by exploring the initial drivers, global, as well as local, that caused the firm to invest in a full-time resource to both create and implement the company's first sustainability policy. It reflects on how those initial drivers have changed over time, yet the key elements of the policy remain untouched and, indeed, in some cases, strengthened.

It reveals the rationale behind the adoption of a specific and somewhat academic methodology for developing the first sustainability policy and details the physical process used to engage with key internal and external stakeholders. This effectively reveals a blue-print for creating a policy and plan, a blue print that could be adopted by any business looking at this same issue. The end result for Auckland Airport was a complex plan, covering 14 areas of business performance and creating 41 targets to be achieved at various times between 1 July 2008 and 30 June 2012. The study covers specific initiatives and shares some of the barriers—political, organisational and human—to their implementation and success. It documents the resistance to, and concerns over, adopting and making public sustainability performance targets. Yet, despite concerns, at its conclusion in June 2012, the company had achieved an impressive 73 per cent success rate and had received national and international recognition for its efforts.

Many lessons were learnt, from the first five years, and these helped shape a more consultative approach to developing a second sustainability policy and plan in 2012. The development process included benchmarking against international best practice, not just in the airports sector, as well as divisional workshops to (re)engage the business. Human, organisational and political factors influenced both the process and the final outcomes, and these are both detailed and discussed. The study goes on to cover, in detail, Auckland Airport's latest plan that sets long-term quantitative targets (out to 2020), for key areas of environmental performance, as well as qualitative targets in the area of supply chain, reporting, urban design and airport community engagement. The study seeks to discover if the change in the sustainability policy wording is pure semantics or the result of a real need and desire to change the business' focus and/or direction.

Auckland Airport has publicly disclosed its new policy and plan and is now developing the mechanisms to manage key areas, report internally and disclose externally performance, over the long term. This, in itself, has raised a number of issues, for the whole business as well as specific divisions, including scope, boundary, key performance areas, base lines and targets. These are being addressed and the study will provide an interesting insight into one company's approach to embedding sustainability into a business with a relatively simple business model but with some complex sustainability issues.

Implementing Sustainable Business Practice at Auckland Airport: Drivers, Responses and Outcomes[4]

Introduction

In 2005, the senior management at Auckland Airport undertook a strategic risk review, looking out to 2015, which covered all manner of business risk including environmental. It included, therefore, the potential fiscal risks associated with an impending domestic emission trading scheme and the introduction of a carbon price into the New Zealand economy. The timing of this review was quite crucial as it occurred at a time when climate change was making media headlines around the globe. The Intergovernmental Panel on Climate Change (IPCC), who advises the United Nations Framework Convention on Climate Change (UNFCCC), had produced, in 1999, a document highlighting the contribution aviation makes to climate change. The IPCC stated that aviation's contribution to climate change, globally, at the turn of the century, was between two and three per cent, but, based on projected growth, was likely to rise to five per cent by 2025.

The European Union (EU) was discussing introducing a carbon tax on airlines flying into European airspace, and domestically the New Zealand Government was formulating a comprehensive domestic emissions trading to cover all six greenhouse gases and all sectors. In addition, climate change was in the public eye globally as Al Gore was undertaking a comprehensive world tour with his 'Inconvenient Truth' multimedia presentation. The company concluded that the number of unknowns around the financial and reputational risk for aviation, tourism and airports was sufficient to warrant the creation of a full-time resource in the area of climate change and business sustainability. The standard business risk assessment process highlighted the issue of climate change and further analysis revealed a complex global sustainability issue that could impact every area of business

4. Author: Martin Fryer PGDip, CEnv (Sustainability and Environmental Manager, Auckland Airport).

development. These impacts ranged from purely economic ones, increasing operating costs from the introduction of a carbon price on energy, fuel and waste, and the requirements of carbon trading, to socioeconomic ones such as a potential down-turn in international passenger numbers from key markets such as Europe and America, where consumers may alter travel habits due to awareness and action regarding their own carbon footprints. For an airport business involved in long-haul international tourism and trade, climate change was the catalyst for the development of business systems that embraced sustainability.

Policy Formulation and Implementation

In September 2007, the company appointed a sustainability advisor tasked with creating the company's first sustainability policy and implementing it into the whole business. The company has a simple structure with three distinct divisions: aeronautical, retail and property, all of whom had begun to embrace the concept of sustainable business practice. The formulation of the policy came about through the creation of a sustainability forum. This was an in-house group of staff members from the three divisions as well as corporate areas of communications, finance, legal and human resources. The group was assembled to try and ensure representation of key stakeholders: staff, community, shareholders, airlines, tenants, etc. Over a period of six months, the forum agreed on 14 key areas of business performance that were seen as essential to include in the sustainability policy (Appendix A). These 14 areas and their associated targets and key performance indicators were chosen following research into current best practice for airports and other New Zealand businesses.

Internal research included reviewing existing structures and personnel currently managing the issues, within Auckland Airport, and the current and potential future levels of engagement, i.e. how likely was it to work and targets be achieved given current levels of awareness, engagement and ownership. Meetings with general managers revealed very different stances on sustainability issues. Climate change was always a key topic of conversation, yet views varied widely from the total sceptics to those that accepted climate change but could not accept an anthropogenic root cause. In an attempt to ensure the sustainability plan was embedded into day-to-day business practice, a clear and easily communicated framework of responsibility and accountability was created and distributed across the business.

This process produced the company's first five-year sustainability action plan with 41 targets to be achieved, at varying points, between the financial years 2008 and 2012. Monitoring of performance was undertaken by the introduction of a simple scoring system. This assigned a score for each target as follows:

Target achieved: 5 points.

Target on track: 3 points.

Target requires work: 2 points.

Target not achieved: 1 point.

The scoring system was designed to foster ownership and engagement; the category 'on track' provided reassurance to the business and business owners that things would be achieved. The category of 'requires work' was a way of communicating that more attention or focus needed to be given to a particular issue or perhaps additional resource needed to be provided.

This enabled an annual scorecard or dashboard to be presented to interested parties (Appendix Table A-1.1) as well as a graphical representation of performance overtime (Appendix Figure A-1.1).

Overall Results and Analysis of Environmental Performance

Overall the company achieved 73 per cent of the 41 targets associated with its first sustainability policy. Targets were achieved across all areas, environmental, social and economic. The company found social targets the easiest to meet with a 100 per cent success rate, environmental second with 60 per cent and economic last at 40 per cent.

The environmental targets were around six key areas:

- Climate change: with absolute and intensity (per passenger) targets.
- Energy and fuel efficiency: with absolute emission reductions and engine type targets.
- Environmental sustainability: with absolute targets for reducing waste to landfill.
- Environmentally sustainable design: with targets for green building accreditations and energy efficiency.

- Surface access: with targets for increasing public transport use and cycling and reducing single occupancy vehicles.
- Water conservation: with absolute and intensity (per passenger) reduction targets.

The company was able to meet its climate change targets ahead of schedule through the implementation of an energy efficiency programme that focused on the key area of HVAC (heating ventilation and air conditioning). Projects included ultraviolet C (UVC) cleaning of coils, installation of variable speed drives, updating building management system, software and optimising plant and equipment.

Targets for reducing emissions from vehicle fleet were also met with removal of company cars and renewal of older fleet vehicles. However the introduction of LEVs (low-emission vehicles) was not monitored, as there were a number of standards for LEVs, and not a single one the company could adopt. This was symptomatic of a lack of engagement by the company's procurement department that had a healthy scepticism about green purchasing policies and their ability to add value and reduce business costs.

Targets for managing waste were met but only within the last 12 months of the five-year plan. The reduction of waste to landfill target was not met. This was one area where responsibility for waste management within the organisation was not clearly defined and existing contractual arrangements had no requirements for collecting and collating waste data. This prevented the introduction of working practices to ensure suitable key performance indicators could be created and targets achieved.

Targets for environmentally sustainable design were met with the exception of engaging property and retail tenants in energy initiatives. Several attempts were made to work with property tenants on energy-efficient lighting projects but there was an insistence by the property division on a shared savings model, so they could recoup the capital investment. Despite a tenant agreeing to this approach the division decided, the administration costs associated with the shared savings model did not make it economically viable. It proved even more difficult to engage retail tenants mainly due to the fact that the focus of the energy-efficiency programme was related to heating and cooling rather than lighting.

Only one of four targets relating to surface access was met. This related to the company's travel plan and the encouragement of airport staff to car-pool to work. This was a key finding from the review of performance. In areas where the airport company had direct control such as the marketing of a carpooling scheme, the results were more easily achieved than where the company could only apply influence such as the provision and uptake of public transport.

Both targets for water conservation were met, however this only occurred in the final 12 months of the plan. Water conservation had taken place in the form of improved leak detection and resolving these issues. The company did not generate a comprehensive water efficiency plan which would have enabled the achievement of this target at a much earlier date.

Creating a New Vision

In late 2011, Auckland Airport turned its attention to a successor to its first sustainability plan. An initial review was undertaken of Auckland Airports approach, against international best practice in the airports sector, as well as associated industries such as retail and property. This highlighted the fact that many corporates were not setting specific targets around key sustainability areas and few were looking long term, i.e. 2030 timelines. The company then produced a consultation document for circulation to, and engagement with, key stakeholders. This was made available, in May 2012, on the company's website and was communicated to stakeholders through various media and methods including:

- Local newspapers (Manukau Courier, Eastern Bays Courier, Western Leader);
- Full page advertorial in Life and Leisure Magazine;
- Airport Times (company monthly newsletter) articles;
- Face-to-face meetings with local Iwi;
- Direct mailing to BARNZ (Board of Airline Representatives of New Zealand) — representing international airlines using Auckland Airport;
- Professional Networks—LinkedIn Groups such as Sustainability Leaders and Airport Sustainability.

The results were collated and reviewed over a period of several months and presented to senior management in October 2012 (Appendix Table A-1.2).

The company had updated its environmental policy in late 2011 (and again in November 2012), introducing, for the first time, the concept of biodiversity which enabled targets to be proposed in this area (Appendix B). The sustainability policy was also revised in November 2012, following the period of stakeholder consultation (Appendix C). Revision of the sustainability policy enabled a rationalisation of the target areas from 14 down to 10. This reduces the complexity of the sustainability plan and eases communication of performance and results whilst continuing to meet stakeholder expectations.

Both policies were again updated and the draft of sustainability policy and plan was refined and agreed by the company's Leadership Team in November 2012. Implementation of the policy is now well underway with specific plans being created around waste minimisation, water minimisation and energy management. Research has been undertaken into creating a base line of carbon emissions from ground transport as a precursor to monitoring and reporting.

Sustainability Reporting

Drivers for reporting and reporting formats have been an on-going topic of discussion at Leadership Team level and the Global Reporting Intiative (GRI) process for determining materiality and reporting priority (identification, prioritisation and validation) has been used to assist the business in producing a transparent report of performance.

Identification

The GRI Airport Operators Sector Supplement (AOSS) identifies the following 10 'material' issues for airports (see later) and these were used as a starting point for discussion:

- Impact on local communities through employment and business development.
- Measurement and monitoring of noise.
- Impacts on biodiversity.

- Management of water and effluents.
- Implications for labour practices related to pandemic diseases.
- Exposure to pollutants and aircraft noise.
- Human trafficking.
- Resettlement of local communities.
- Assessing business continuity.
- Emergency preparedness.

It was decided not all the AOSS issues were material for Auckland Airport. The company then turned to the 2012 stakeholder engagement exercise conducted as part of the creation of the company's second sustainability action plan. Seven issues were identified as material by stakeholders and a further five by the business (included in the company's current sustainability policy).

- Noise.
- Energy and fuel (carbon).
- Transport.
- Community and Iwi relations.
- Social contribution.
- Biodiversity.
- Waste management.
- Water conservation.
- Best practice urban design.
- Culture, heritage and creativity.
- Airport community and passenger engagement.
- Supply chain engagement.

Prioritisation

Applying a simple scoring system, based on the quantum of consultation responses, allows the materiality of these issues to be plotted to enable the priority for reporting to be determined.

Figure 1.1

Prioritisation

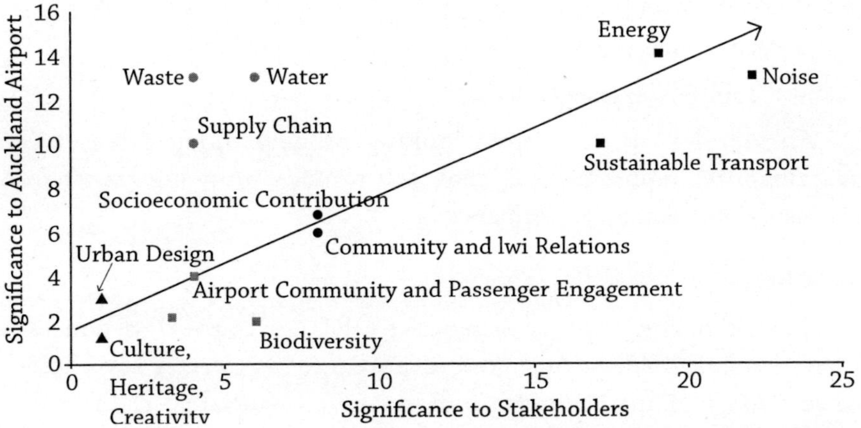

In the case of Auckland Airport, the issues of aircraft noise, energy usage (carbon) and transport are seen as material by both the business and stakeholders. Equal weighting, but lower priority is given to community-Iwi relations and socioeconomic contribution.

Interestingly, the business puts far more emphasis on water conservation, waste minimisation and supply chain engagement. This is to be expected as the business would be far more aware of the potential economic benefits of paying attention to these issues. At the lower end of the reporting priority scale are airport community and passenger engagement as well as biodiversity. These could be reported periodically, as and when initiatives are completed. Culture, heritage and creativity and best practice urban design have been identified by the business as important, to the sustainable development of the airport, but do not potentially require specific coverage in sustainability reports.

This revealed eight material issues that will form the basis of annual long-term sustainability reporting for Auckland Airport:

- Noise.
- Energy and fuel (carbon).
- Transport.

- Community and Iwi relations.
- Social/economic contribution.
- Waste management.
- Water conservation.
- Supply chain engagement.

Auckland Airport is now looking at developing a web-based sustainability reporting tool that will provide on-going transparent disclosure of company performance.

Validation

The GRI materiality process requires validation of the chosen material issues by stakeholders. Auckland Airport intends to refer stakeholders to the sustainability reporting tool in its FY(financial year) 14 Interim Report and request feedback prior to disclosing full FY14 sustainability performance at financial year end.

Conclusions

Auckland Airport took a comprehensive if somewhat academic approach to implementing its first sustainability policy and plan. The approach proved incredibly effective at a time when knowledge and understanding amongst stakeholders was far more limited than it is today. It was important therefore for the company to provide leadership in the sustainability space, and sound governance structures have supported engagement across the entire business. The result is a successful first five years with 73 per cent of sustainability targets being achieved.

It is interesting to note both the sustainability and environmental policies have been redrafted and together cover stakeholder expectations for environmental performance. The company maintains separate policies to show the equal importance of sustainable business performance and environmental performance despite clear overlaps in many areas. The engagement of stakeholders has differed over time, with the initial consultation in 2007 being very limited, and policy and plan formulation essentially being an in-house process. The second period of consultation in 2012 reflects the increased awareness and understanding of sustainability issues across all stakeholder groups, and the use of differing media has enabled much wider stakeholder involvement.

The drivers for sustainability disclosures have also changed since 2005 with aircraft noise being a far more material issue for Auckland Airport than climate change. The current sustainability plan with targets out to 2020 and aspirational goals out to 2030 will maintain Auckland Airports' leadership position in this space. Quantitative targets will add value to the business in terms of returns to the bottom line as well as influencing key stakeholders within its supply chain. Qualitative targets will ensure it continues to meet its obligations as a good corporate citizen in the environmental, social and cultural areas. Auckland Airport's approach to reporting is following international best practice by using GRI methods and tools. Development of a web-based reporting tool for the long term, that includes disclosure of a best practice methodology, reflects the company's commitment to transparency.

Recommendations

Companies at any stage of their sustainability journey would do well to look at Auckland Airport as a case study. It provides an opportunity to learn some important lessons around embedding sustainable business practice through: stakeholder engagement, policy formulation and implementation, target setting, performance monitoring and transparent disclosure. It is recommended that further research be undertaken into the airport company's approach to embedding sustainable business practice and the success of its 2020 vision.

Appendix A

Company Policy: Sustainability

Auckland Airport's core business is the operation of New Zealand's largest international and domestic airport. We are proud of our airport and are committed to providing a unique and compelling customer experience, where every employee represents our country. Sustainability is integral to the way we think, the way we conduct ourselves, and the approach we take to our day-to-day business.

In particular Auckland Airport will:

- Operate in a safe, secure, sustainable and efficient manner.
- Balance the economic, environmental, social and cultural aspects of all our activities through a proactive sustainability programme.
- Targets for the following key areas as practicable.
    1. Climate change.
    2. Community and sponsorship.
    3. Culture and heritage.
    4. Economic performance.
    5. Energy and fuel efficiency.
    6. Environmental sustainability.
    7. Environmentally sustainable design (buildings and infrastructure).
    8. Operational efficiency.
    9. Our People (Our Spirit).
    10. Procurement.
    11. Reporting.
    12. Safety.
    13. Surface access.
    14. Water conservation.
- Work closely with all stakeholders to enhance the sustainability of New Zealand as a place to live and work, to trade, and to host local and international tourists.

- Communicate performance to relevant stakeholders on a timely basis.

Successful implementation of this policy will enhance Auckland Airport's reputation, customer loyalty, passenger experience and create long-term value for our owners, as well as our other stakeholders.

Signed:

Dated: 19 August 2008

Table A-1.1

Auckland Airport Sustainability Performance Dashboard FY08 to FY12

	Society	*FY08*	*FY09*	*FY10*	*FY11*	*FY12*
1.0	Culture and Heritage					
1.1	Include physical representations of history and culture in the passanger experience	5	5	5	5	5
1.2	Enhance existing heritage areas (ongoing)	0	2	5	5	5
1.3	Develop (where appropriate) new heritage areas (ongoing)	0	1	2	2	5
1.4	Produce an airport management report (annual)	5	5	5	5	5
1.5	Consult local lwi on all relevant developments	5	5	5	5	5
2.0	Society					
2.1	Investigate and implement a staff scheme to support voluntary work FY12	0	2	5	5	5
2.2	Maintain two AACT trustees (ongoing)	5	5	5	5	5
2.3	Review sponsorship spend and recipients by FY09	2	5	5	5	5
2.4	Identify and target key recipients by FY12	0	2	5	5	5
3.0	Our People					
3.1	Identified responsibilities under sustainibility policy by FY10.	2	2	5	5	5
3.2	Ensure the PDS system recognises individual performance in sustainibility (ongoing)	0	2	5	5	5
4.0	Reporting					
4.1	Produce a CEO sustainability performance report (quaterly)	0	2	5	5	5
4.2	Complete a CDP return (annual)	5	5	5	5	5
4.3	Ensure reporting of sustainability performance is web based by FY10	2	2	5	5	5

contd...

...contd...

4.4	Update sustainability performance online annually	0	2	5	5	5
5.0	Safety					
5.1	Zero LTI by FY12	2	2	2	5	5
5.2	Zero LTIFR by FY12	2	2	2	5	5
	Economy	*FY08*	*FY09*	*FY10*	*FY11*	*FY12*
1.0	Operational efficiency					
1.1	CAA risk rating maintained as low (ongoing)	5	5	5	5	5
1.2	Establish operational efficiency targets and appropriate KPI by FY09	0	0	0	0	0
2.0	Economic performance					
2.1	TSR to increase year on year by FY12	0	5	5	0	0
2.2	Contribution to regional GDP to be maintained and measured (FY11, FY16, etc.)	2	2	5	5	5
3.0	Procurement					
3.1	Purchasing policy and guidelines updated to include sustainability criteria by FY10	0	2	5	5	5
3.2	Develop suitable KPI and guidelines for sustainable procurement by FY10	0	2	0	0	0
3.3	Sustainability of supply chain and key suppliers determined by FY11	0	2	2	0	0
3.4	Increase purchase of sustainable goods and service from FY11	0	2	2	1	0
	Environment	*FY08*	*FY09*	*FY10*	*FY11*	*FY12*
1.0	Climate change					
1.1	Reduce CO_2 emission per international passenger by 10% by FY12	2	1	5	5	5
1.2	Reduce the total green house gas CO_2 emissions footprint of Auckland Airport by 5% by FY12	1	1	2	5	5
2.0	Energy and fuel efficiency					
2.1	Increase number of LEV vehicles in company fleet by 10% by FY12	2	1	1	1	0
2.2	Vehicle fleet CO_2 emissions to reduce by 10% by FY12	2	2	5	5	5
3.0	Environmental sustainability					
3.1	Understand the quantities, type and true cost of disposal of waste by FY10	0	2	0	0	5

contd...

...contd...

3.2	Reduce the quantity of waste going to landfill by 10% by FY12	1	1	2	1	0
4.0	Environmentally sustainable design					
4.1	Pier B extension on GBC accredited by end FY09	2	0	5	5	5
4.2	Retail and property intensive energy users involved in energy efficiency by FY09,	0	0	0	0	0
4.3	Property specific sustainability initiatives finalised by FY09	2	5	5	5	5
4.4	Selected commercial developments Green Star rated by FY12	2	2	2	2	5
5.0	Surface access					
5.1	Public transport mode share to increase by 10% by FY12	2	2	2	1	0
5.2	Multiple occupancy vehicle rate to increase by 10% by FY12	2	2	2	1	0
5.3	Registered lift carpoolers to increase by 10% per annum by FY12	5	5	5	5	5
5.4	Number cycling to increase by 10% by FY12	0	1	1	1	0
6.0	Water conservation					
6.1	Decrease potable water use per passenger by 10% by FY12	2	1	1	1	5
6.2	Decrease total potable water use by 5% by FY12	2	1	1	1	5

Note: AACT : Auckland Airport Community Trust; PDS: Public Distribution Center; LTI: Lost Time Incidents; LTIFR: Lost Time Injury Frequency Rate; CAA: Civil Aviati on Authority of New Zealand; KPI: Key Performance Indicator; TSR: Total Shareholder Return.

Key	
Achieved	5
On Track	2
Work Required	1
Not Acheived	0

Figure A-1.1

Auckland Airport Sustainability Performance Summary (2008-2012)

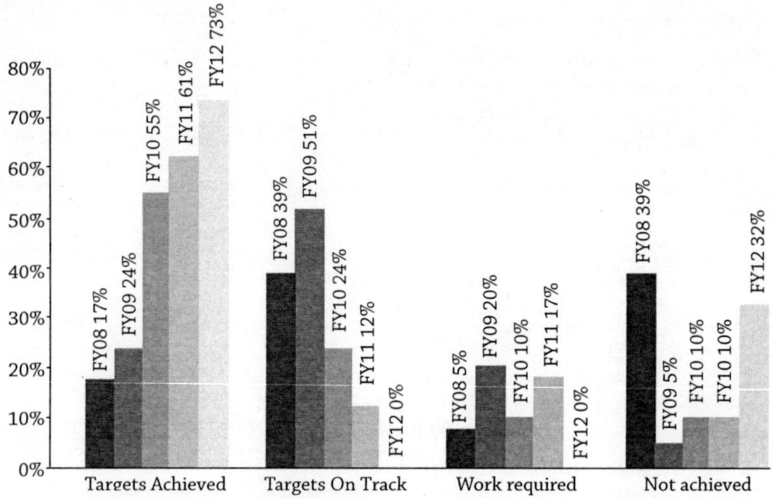

Table A-1.2

Auckland Airport 2030 Consultation Feedback 2012

Respondent	*Source*	*Aspect*	*Comment*	*Note*
AA staff	Internal staff workshops	Timeline	Push out to 2040 to align with Auckland City Plan	Makes sense in some ways but surely private sector should be able to achieve things ahead of local government?
AA staff	Internal staff workshops	Key areas	Extend to 16 to include other emissions and product	Would align us with GRI reporting but is that the key driver? Do we not have an opportunity to further simplify whilst still meeting stakeholder expectation?
AA staff	Internal staff workshops	General	Plan will require review and evolution over the period	Agree and we need to show how this will happen and how stakeholders can contribute going forwards

contd...

...contd...

Individual	NZ Life & Leisure	Social	Should AA offer scholarships in sustainability, training of local youth to ensure next generation understand the sustainability issue	Agree—and we have the relationship with MIT are we promothing this as a sustainability initiative?
Bus and Coach Assoc.	AT working group	Transport	Support increase in multi occupancy vehicle use and PT Support the idea of a bus rapid transit system to CBD.	Agree—we need to reinvigorate the travel plan and improve PT offerings with AT
Individual	Manukau Courier	General	More use of solar panels, creat a solar tree, plant more native species around the precinct, recycle more waste, phyrolysis of plastic for fuel production, more futuristic use of architecture, an aviation museum, cinema and a few rides for pax to pass the time.	Agree—biodiversity element of environmental policy, landscape overlay and good urban design should achieve these outcomes. Solar energy is being reviewed and improving waste recovery and recycling is well underway.
Individual	Manukau Courier	Aircraft noise	Respondent unaware of noise management practice.	
Individual	Eastern courier	Aircraft noise	Respondent interested to know if planes could use a different take-off procedure when the wind is from the East	
Individual	Manukau Courier	Aircraft noise	Respondent believes aircraft noise exceeds Ldn 60 dBA outside the Mana	To be shared with the ANCCG.
Individual	Eastern courier	Aircraft noise	Respondent suggested restrictions on visual approaches at night and departures should have a specific geograhpical turning point away from residential areas.	
Individual	Eastern Courier	Aircraft noise	Introduce a night curfew for departures to the East.	

contd...

...contd...

Individual	Website	Noise and fuel	Incentivise airlines that fly fuel efficient and quieter aircraft	Interesting concept and could be discussed in relation to mandating GPU/PCA usage?
Individual	Manukau Courier	Energy	Convert vehicle fleet to run on methane gas and source from Mangere STW.	Interesting idea?
Individual	Manukau Courier	Energy	Use more solar panel/solar water heating	Agree—economics has changed markedly in the past six months. Need to review in terms of airport infrastructure and property developments
Linkedin Networks: Sustainability Leaders and Airport Sustainability	Comments from:	Areas		Some interesting idea and concepts many of which are already on our wish list.
	Texas	Energy		
	Dallas Fort Worth	Climate change		
	India	Biofuels		
	American Airlines	Cycling		
	Turkey	PT		
	New Zealand	Sustainable transport		
		Cycling		
		Economic and community health and well being		

Note: GRI: Global Reporting Initiative; AA: Auckland Airport; AT: Auckland Tranport; PT: Public transport; MIT: Manukau Institute of Technology; GPU: Ground Power Unit; PCA: Pre-conditioned air.

Appendix B

Environment Company Policy

Purpose

To protect the environment and minimise the impact, of airport operations, on the local community. This policy enables regulatory risk to be reduced and the company to contribute to maintaining and enhancing New Zealand's clean, green image.

As the operator of New Zealand's largest international airport and the key gateway into New Zealand, Auckland Airport is acutely aware of the need to manage environmental risk. Auckland Airport is located on the very edge of the Manukau Harbour. It is a unique site that includes sensitive natural environments, rich local heritage and culture, but is also close to residential communities.

In order to manage the environmental risks, associated with its operations, Auckland Airport shall:

- Conduct its operations in compliance with all applicable environmental legislation, plans, standards and codes of practice.
- Strive to achieve best practice in environmental management relative to other international airports.
- Strive to prevent pollution by identifying, controlling and minimising adverse environmental impacts.
- Protect and, wherever possible, enhance habitats and their associated wildlife, without compromising aircraft safety.
- Encourage a commitment to the environment from all stakeholders, including: employees, tenants, customers and the local community.
- Maintain a documented management system that includes the setting of objectives and targets, training, audit, management review and the requirement for continual improvement.
- Report its performance to interested stakeholders on a regular basis.

Appendix C

Sustainability: Company Policy

Purpose

To ensure the company develops in a sustainable manner. That is, development that meets the needs of the current generation, without compromising the ability, of future generations, to meet their needs.

To Auckland Airport sustainability means operating in a manner that finds the right balance between the economic, environmental, social and cultural aspects of all our activities.

Sustainability is part of business as usual, as well as an essential element in creating and realising the long-term vision for Auckland Airport.

Auckland Airport will:

- Implement a proactive sustainability programme.
- Produce sustainability metrics, set specific, achievable sustainability goals and targets for the following key areas:
  1. Energy and fuel efficiency
  2. Water conservation
  3. Waste minimisation
  4. Environmental protection and enhancement
  5. Sustainable transport
  6. Best practice urban design
  7. Culture, heritage and creativity
  8. Airport community and passenger engagement
  9. Supply chain engagement
  10. Reporting performance
- Work closely with all our stakeholders to enhance the sustainability of New Zealand as a place to live and work, to trade and to host local and international tourists.

Successful implementation of this policy will enhance Auckland Airport's reputation, customer loyalty, passenger experience, and create long term value for our owners, as well as our other stakeholders.

CSR and Corporate Governance: A Case of India and New Zealand

In 2013, India was the first (and remains the only) country to introduce a statutory requirement on business to engage in CSR, providing a unique setting for researching CSR implementation.

Over the last decade, Indian businesses were recognising the challenge of going global and also operating nationally with evolving regulatory requirements around environmental and social performances. There were also many players like the regulators, the business chambers, enthusiastic civil societies who were pushing the agenda of sustainability through diverse mandatory and voluntary requirements and focusing on various issues of pollution, energy, climate change safety, labour practices. Sustainability reporting was initiated 10 years back by small percentage of companies steadily growing to a reasonable number in 2011 (Global Reporting Initiative [GRI], 2013).

In the financial sector, the Reserve Bank of India (RBI) had drawn the attention of banks to their role in CSR, sustainable development and non-financial reporting in its circular dated December 2007. The Ministry of Corporate Affairs, Government of India issued the National Voluntary Guidelines on Social, Environmental and Economic Responsibilities of Business on 8 July 2011. The guidelines provide a robust framework that may be adopted voluntarily by companies to address interests of various stakeholders, including employees, customers and investors. The guidelines also suggests the central public sector enterprises (CPSEs) to include the non-commercial target, i.e. CSR spend along with their financial targets. In most recent times, the Securities and Exchange Board of India (SEBI) mandated from November 2011 that the 100 top listed entities must submit Business Responsibility Reports, as a part of their annual reports.

Above all the newly enacted Companies Act, 2013 made it mandatory to include in the board's report a CSR policy and to spend two per cent of average net profit of last three years on CSR activities, and had set a clear forward movement on the agenda. According to the Indian Institute of Corporate Affairs, this will result in a minimum of 6,000 Indian companies (including small and medium enterprises [SMEs]) being required to undertake CSR projects. Such initiatives will be new for many corporates and SMEs, and some estimates indicate that the CSR commitments could amount to as much as ₹20,000 crores (equal to 4 billion New Zealand dollars).

Recently, Collins *et al.* (2013) surveyed an exploratory sample of companies in India for a cross-country comparison with New Zealand business practices. This research builds on a longitudinal study of business sustainability practices in New Zealand. The research extends the New Zealand business practices survey to an exploratory sample of companies in India so that cross-country comparisons can be made. The key research question is: what is the nature and extent of business adoption of sustainability practices in New Zealand and India?

The survey in India was exploratory, covering the top performing CPSEs which are owned by Government of India (GoI). Out of the total 239 CPSEs, the GoI identified 89 CPSEs to ensure financial autonomy based on their cognitive activities and performance and classified them into three categories: Maharatna, Navratna and Miniratna. The government had nominated these as being of strategic importance and having the potential to emerge as global players (Locke and Geeta, 2013). The data were from a survey of 89 corporates surveyed in 2013 with a 51 per cent response rate, for a total of 45 participants. The survey was distributed to participants (corporates) attending a Boards Conclave jointly organised by International Management Institute and DPE, Ministry of Heavy Industries and Public Enterprises at Bangalore, India, in June 2013 and to the participants attending the corporate executive development programme organised by International Management Institute, New Delhi, in August 2013. The respondents were either directors or senior executives.

Firm size matters to the survey results because it is one of the best predictors of the uptake of sustainability strategies as research has shown that smaller firms are less likely to adopt social or environmental practices (Lawrence *et al.*, 2006). Because all of the Indian businesses surveyed were large, the sample used from the New Zealand survey was only the large businesses. Out of 520 respondents, 94 companies (or 18% of respondents) were large businesses, defined in this study as over 100 employees.

A key limitation to the study is that the sample size from both countries is small. Statistical analysis has not been conducted as we are not claiming country representativeness or generalisability, rather the surveys are a case study of a sample of companies from both countries. The language in the results carefully uses 'surveyed companies' rather than claiming to speak for companies in New Zealand and India. Our intention follows Yin (2013), that is, for theoretical generalisability, testing how our results are similar or are different or can extend the literature on the drivers and barriers related to the uptake of sustainability practices.

The following section details the results of the surveys in both countries.

Results

Sustainability Practices: The findings indicate differences in the CSR/sustainability approaches and practices of Indian corporates and New Zealand corporates. For instance, a recycling programme was the most common environmental practice for 95 per cent New Zealand companies, whereas only 20 per cent per cent of the surveyed Indian companies recycled.

Drivers and Barriers: Likewise, when the drivers to adopt sustainability strategies were examined, 67 per cent of companies in India (compared with 36% in New Zealand) were more motivated by government regulations, while New Zealand companies were more likely to be motivated by reputation and brand (New Zealand: 80%; India: 42%).

Cost Management: With regard to cost management and/or cost reduction, 64 per cent of the surveyed New Zealand companies were more likely to see the business case for adopting sustainability as a driver, but only 16 per cent Indian companies cited costs as a driver. Overall, the New Zealand companies surveyed were more engaged in both environmental and social practices compared to Indian companies.

Overall, the New Zealand companies surveyed were more engaged in both environmental and social practices compared to surveyed companies in India. These results are perhaps not surprising as we would expect to see more practices in a developed country. Although interestingly, nearly 20 per cent of surveyed New Zealand companies had no environmental sustainability practices at all. There were no Indian businesses in that category. Surveyed companies in India also were much more likely to issue an environmental or sustainability report. Previous research has shown that New Zealand, compared to international standards, has a very low reporting rate (Collins *et al.*, 2010).

Figure 1.2

Environmental Practices in New Zealand and India

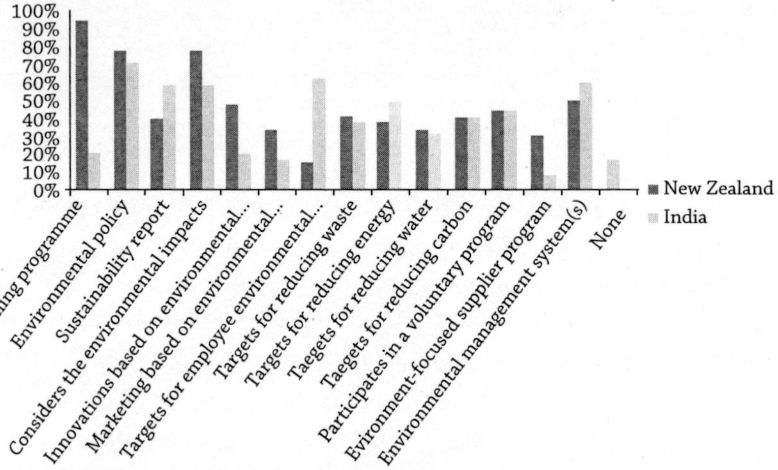

Note: The following is a complete list of environmental practices a respondent could choose from: has a recycling program; has a company environmental policy statement; produces a public environmental and/or sustainability report; develops product and service innovations based on environmental benefits; considers the environmental impact of our products, processes and/or services; marketing or image based on environmental claims; has measurable targets for employee training programmes related to our environmental goals; has measurable targets for reducing waste; has measurable targets for reducing energy; has measurable targets for reducing water; has measurable targets for reducing carbon; participates in a voluntary environmental programme; has an environment-focused supplier programme; has environmental management system(s).

In both countries, surveyed respondents were more engaged in social rather than environmental sustainability. The results relate to the literature on trade-offs. Social and environmental sustainability were not treated equally, but rather social sustainability practices were much more likely to be adopted. Why social sustainability practices are more common for business is an area that needs further research.

The findings on drivers align with Mishra *et al.* (2012) showing that the business case for sustainability has not found much traction in India. The complex relationship of costs and sustainability was one of the key findings of the study. Costs were simultaneously a driver and a barrier, and impacted the business case for sustainability.

Clearly the recently passed Companies Act (2013) will change the sustainability landscape in India. For surveyed companies in India, government regulation was the key driver compared to reputation and brand for surveyed New Zealand companies. What will have to be left for future study is whether the different drivers lead to a divergence in practices between the two countries. Is there a difference in what practices a business chooses to implement because of regulation versus what it chooses to do voluntarily?

Figure 1.3

Social Practices in New Zealand and India

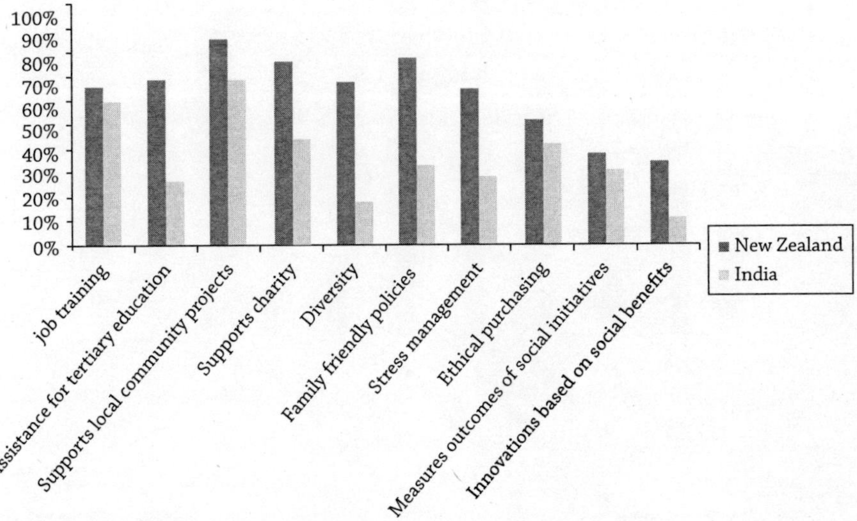

Note: The following is a complete list of social practices a respondent could choose from: provides job training; provides assistance for employees to obtain tertiary education; gives time, money, products or services to local community projects; contributes time, money, products or services to charity; considers diversity in hiring decisions; has family friendly policies; has stress management initiatives; has ethical purchasing policies; measures outcomes/impacts of socially related initiatives; develops product and service innovations based on social benefits.

Figure 1.4

Drivers to Adopting Sustainability Practices in New Zealand and India

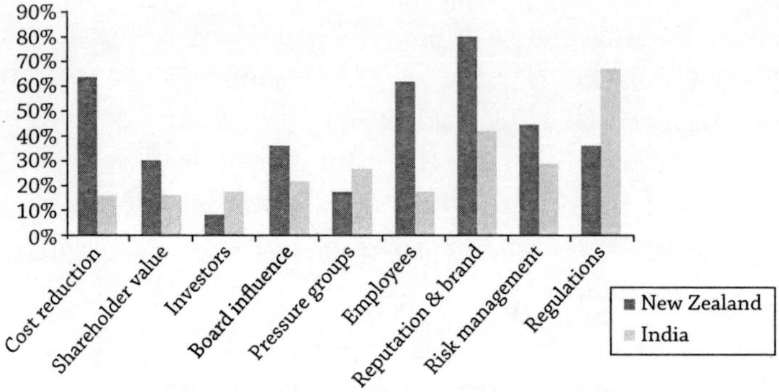

Note: The following is a complete list of drivers a respondent could choose from: cost management/reduction; improved shareholder value; investor pressure, including socially responsible investing; board influence; outside pressure groups; attractiveness to employees; reputation and brand; risk management; government regulations; other.

Figure 1.5

Barriers to Adopting Sustainability Practices in New Zealand and India

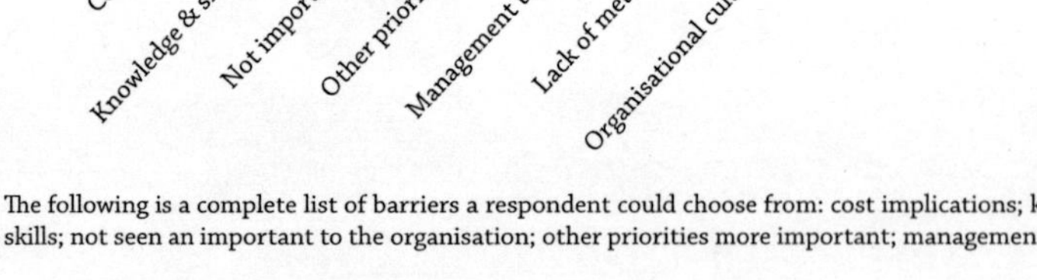

Note: The following is a complete list of barriers a respondent could choose from: cost implications; knowledge and skills; not seen an important to the organisation; other priorities more important; management time; other.

References

Collins, Eva, Geeta Duppati and A.K. Rath (2013). "Sustainability Practices, Drivers and Barriers: A New Zealand/India Comparison", *Journal of Commerce & Business Studies* 1.

——— (2013). *Corporate Social Responsibility and Sustainability Practices: A New Zealand/India Comparison.* Te Kaarohirohi (PRME) conference, Waikato-Tainui College. November 19-20.

2

Board Structure and Duality

Theoretical Perspective

This chapter presents the conceptual framework of the corporate governance, corporate boards and their structures, and other related issues like duality, independent directors and so on.

About Corporate Governance

The evolution of governance can be traced back to the work of Berle and Means (1932). They suggested that the rise of the modern corporation has brought a concentration of economic power which can compete on equal terms with the modern state. Where its own interests are concerned, it even attempts to dominate the state. The future may see the economic organism, now typified by the corporation, not only on an equal plane with the state, but possibly even superseding it as the dominant form of social organisation.

In fact, Berle and Means' concern about the separation of ownership from control was not only about managers' lack of accountability to investors. It was also a concern about managers' lack of accountability to society in general.

Board structure is a key determinant of board effectiveness as per the general theory of the 'Dynamics of Board' propounded by Peter Ducker and Michael Porter. Board structure is taken as a synonym for board composition. As a matter of fact, they are two different terms. Board structure is a broader term encompassing board composition, policy regarding board mix, board tenures and other related issues.

However, board composition is the most important component of board structure.

Board Structure and Board Effectiveness

In recent times, one of the primary approaches to improve governance effectiveness is changing the board structure. These changes often focused on reducing the size of the board, and removing or reducing geographic parameters in board composition. In addition, organisations focused on governance models and the competencies of board members. Board structure is a foundation for an effective board, focusing on the background interests, affiliations and position of members. It is concerned with balances of power and is at the heart of board performance and accountability. Board structure distinguishes between those directors who hold management positions in the company and those who do not. Those with management positions are referred as executive directors and those without are referred as non-executive directors (NEDs) in UK and outside the US. Not all NEDs are the same. There is a difference between a NED and an independent NED. 'Independent' directors are defined in the *Cadbury Report* as persons who "apart from directors fees and shareholdings [are] independent of the management and free from any business or other relationships which could materially interfere with the exercise of the independent judgment."

The Board and Management Differentiated

The responsibility of the board is to govern the company, while the manager's task is to run its business. According to Ticker, the work of a board is to deal with corporate governance while that of the executive organisation is management.

The two-tier board in which there are no common members between the board and the executive management team, is a model more popular in Germany.

Basically there are four alternative board structures:

independence, with the goal of increasing the number of directors who are free from conflicts (from the context of Indian independent directors see the newly enacted Companies Act, 2013).

Independent Director: New Zealand Context

Independence of mind is a basic requirement for directors. Each should endeavour to have an independent perspective when making judgments and decisions on matters before the board. This means a director puts the interests of the entity ahead of all other interests, including any separate management interests and those of individual shareholders (except as permitted by law). Directors with an independent perspective are more likely to constructively challenge each other and executives—and thereby increase the board's effectiveness.

Securities and Exchange Commission (SEC) views that there would be practical constraints in New Zealand if too high a level of formal independence is required of boards. With New Zealand's relatively small pool of qualified and experienced directors, there would be a risk in seeking independence at the cost of all. It will lead to missed opportunities to appoint directors who can contribute to the success of entities. SEC considers the underlying issues relating to director independence which can be addressed by:

- Directors having an independent perspective in their decision-making.
- A non-executive director NED being formally classified as independent only where he or she does not represent a substantial shareholder and where the board is satisfied that he or she has no other direct or indirect interest or relationship that could reasonably influence their judgement and decision-making as a director.
- The chairperson of a publicly owned entity being independent.
- The representation of the board including independent director's for every issue.

- Boards of publicly owned entities comprising:
  - a majority of NEDs and
  - a minimum one-third of independent directors.
- Boards taking care to meet all disclosure obligations concerning directors and their interests, including information about the directors, and identify which directors are independent.

Non-Executive Director

The Role of a NED

UK context: Essentially the NED's role is to provide a creative contribution to the board by providing objective criticism. The 1992 Cadbury Report initiated a debate about the main functions and responsibilities of NEDs. Today, it is widely accepted that NEDs have an important contribution to make to the proper running of companies and, therefore, more widely to the economy at large. As the Cadbury Report said: they "should bring an independent judgement to bear on issues of strategy, performance and resources including key appointments and standards of conduct." There is no legal distinction between executive and NEDs. As a consequence, in the UK unitary board structure, NEDs have the same legal duties, responsibilities and potential liabilities as in the growing number of private companies, including executive counterparts. Clearly, it is appreciated that NEDs cannot give the same continuous attention to the business of the company. However, it is important that they show the same commitment to its success as their executive colleagues.

All directors should be capable of seeing company and business issues in a broad perspective. Nonetheless, NEDs are usually chosen because they have a breadth of experience, are of an appropriate calibre and have particular personal qualities. Additionally, they may have some specialist expertise or, perhaps, key contacts in related industries or the city. The Cadbury Committee (1992), Hampel Committee (1998) and Higgs Committee (2003) Reports, some of whose recommendations are included in the revised Combined Code, stress that the board should include independent NEDs of sufficient calibre and number for their views to carry significant weight in the board's deliberations.

The Combined Code, which effectively codifies the main features of the Cadbury, Hampel and Higgs Reports for listed companies, advises that the balance of executive and NEDs should be such that no individual or small group of individuals can dominate the board's decision making. NEDs should comprise not less than half the board. While much of the comment and discussion on NEDs tends to focus on listed companies, it is important to note that they can also make a valuable, albeit somewhat different, contribution to private companies. Indeed, there are a growing number of private companies, including relatively small ones, that are now actively searching for the 'right' NED.

NEDs are expected to focus on board matters and not stray into 'executive direction,' thus providing an independent view of the company that is removed from its day-to-day running. Chairmen and chief executives should use their NEDs to provide general counsel and a different perspective on matters of concern. They should also seek their guidance on particular issues before they are raised at board meetings. Indeed, some of the main specialist roles of a NED will be carried out in a board subcommittee, especially in listed companies.

Key Responsibilities of NEDs

Strategic direction: as an 'outsider,' the NED may have a clearer or wider view of external factors affecting the company and its business environment than the executive directors. The normal role of the NED in strategy formation is therefore to provide a creative and informed contribution and to act as a constructive critic in looking at the objectives and plans devised by the MD and his or her executive team.

- Monitoring: NEDs should take responsibility for monitoring the performance of executive management, especially with regard to the progress made towards achieving the determined company strategy and objectives.
- Communication: The company's and board's effectiveness can benefit from outside contacts and opinions. An important function for NEDs, therefore, can be to help connect the business and board with networks of potentially useful people and organisations. In some cases, the NED will be called upon to represent the company externally.
- Audit: It is the duty of the whole board to ensure that the company accounts properly to its shareholders by presenting a true and fair reflection of its actions and financial performance and that the necessary internal control systems are put into place and monitored regularly and rigorously. NEDs have an important part to play in fulfilling this responsibility, whether or not a formal audit committee (composed of NEDs) of the board has been constituted.

Seven General Duties of Directors: Institute of Directors (IoD), UK

- To act within the powers of the company.
- To promote the success of the company for the benefit of its members as a whole, paying due regard in decision-making to:
  - likely long-term consequences;
  - employees' interests;
  - the need to foster relationships with suppliers, customers and others;
  - the impact of operations on the community and the environment;\
  - the need to maintain high standards of business conduct and to act fairly between members of the company.
- The need to exercise independent judgement.
- The need to exercise reasonable care, skill and diligence.
- The need to avoid conflicts of interest.
- The need to not accept benefits from third parties.
- The need to declare, where applicable, any interest in a transaction or arrangement with the company.

Duality

It is strongly suggested in the literature that separating the post of CEO and chairperson will not only improve firm performance but also help in minimising agency costs. However according to Fama and Jensen (1983), it is proposed by the agency theory that in order to minimise agency cost the structure of leadership could not be separated. It was found that CEO/Chair duality does not impact agency costs. Mcknight and Weir (2009) in a study on UK public listed firms also finds that duality does not play an important role in mitigating agency cost in UK public listed firms.

From an international perspective, the CEO duality has been the dominant board leadership structure of US corporations, in which 70-80 per cent of them combine the roles of CEO and chairperson (Rechner and Dalton, 1991). However, the prevalent corporate governance practice in Europe separates the CEO and chairperson, while only 10 per cent of UK publicly-listed companies combine these two roles (Ang *et al.*, 2000; Higgs, 2003).

On the other side of the world, the board leadership structure of Asian companies lies in the middle of these two extremes. Hong Kong, a former British colony, has a well-developed regulatory framework and capital market. The proportion of CEO duality for public companies in Hong Kong was 54 per cent in 1996 (GuI and Leung, 2004) and 52 per cent from 1995 to 1998 (Chen *et al.*, 2005).

Board leadership structure is one of the most controversial topics in corporate governance literature. Those who advocate either combining or splitting the roles of CEO and chairperson have their own theoretical justifications. However, the terms that express the combining or splitting of the CEO/chairperson are sometimes confusing. Some refer to this as CEO duality and non-duality (Finkelstein and D'Aveni, 1994; Baliga *et al.*, 1996), some as a unitary and dual leadership structure, some as a CEO duality and independent chair (board leadership) structure (Rechner and Dalton, 1991; Boyd, 1995; Rhoades *et al.*, 2001), and some as a joint CEO/

chairperson and separate chairperson (Daily and Dalton, 1997a; 1997b), among other terms.

There are two different views on board leadership structure. Agency theorists argue that the roles of CEO and chairperson should be separated; otherwise, a single person holding both roles will dominate the board. Combining the roles of CEO and chairperson (CEO duality) is generally a sign of a dominant CEO and could render the board ineffective in monitoring managerial opportunism (Daily and Dalton, 1993; Jensen, 1993). As a result, CEO duality enhances CEO entrenchment and reduces board independence (Finkelstein and D'Aveni, 1994; Rhoades *et al.*, 2001), so that less independent directors will be appointed to the board.

Higgs (2003: 23) recommends separating these two roles, which "avoids concentration of authority and power in one individual and differentiates leadership on the board from running of the business." The CEO is responsible for executing the company's policies and running the company, while the chairperson of the board is responsible for running the board and monitoring and evaluating managerial activities.

The board is also responsible for overseeing the process of hiring, firing, evaluating and compensating the CEO, and thus the chairperson should preferably not be the same person whose performance is being assessed; otherwise, one is self-evaluating (Jensen, 1993: 36). CEOs would therefore gain much power "to further their own interests rather than the interests of shareholders" (Weisbach, 1988: 435). Therefore, CEO duality is expected to affect firm performance negatively.

In contrast, stewardship theorists counter-argue that managers are inherently trustworthy and are good stewards of company resources (Donaldson, 1990; Donaldson and Davis, 1991; 1994). Stewardship theorists reckon managers to be self-actualising persons rather than opportunistic shirkers. In addition to financial incentives, managers are also motivated by non-financial motives such as job satisfaction, advancement and recognition, respect for authority, and work ethic. They seek intrinsic satisfaction by performing challenging work. Managers would not risk their career and reputation to do something against the owners' interests.

The future fortunes of managers in terms of employment and pension rights are bound to their current firms, and thus their interests are aligned with the owners'. The supporters of stewardship theory thus advocate that the interests of managers are aligned with those of the owners, and there is no inherent conflict of interest between the two. They believe CEO duality could promote a unified and strong leadership with a clear sense of strategic direction. Because the CEO has knowledge of the business and industry and knows how to run the company, combining these two roles can help in making timely and optimal decisions (Brickley *et al.*, 1997). Thus, CEO duality is expected to affect firm performance positively.

Empirically, the evidence on the duality-performance relationship is still mixed and inconclusive (see, e.g., Daily and Dalton, 1997a; 1997b; Dalton *et al.*, 1999; Kang and Zardkoohi, 2005). Some prior empirical tests support the separation of the roles of CEO and chairperson (see Rechner and Dalton, 1991; Daily and Dalton, 1997a; 1997b; Chen *et al.*, 2005), while others suggest that combining these two roles is preferable (see Donaldson and Davis, 1991; Brickley *et al.*, 1997; Coles *et al.*, 2001). Additional studies do not detect any significant relationship between board leadership structure and firm performance (see, e.g., Baliga *et al.*, 1996; Daily and Dalton, 1992; 1993; 1997a; 1997b; Dalton *et al.*, 1998; Dulewicz and Herbert, 2004).

However, empirical evidence does show that there is no universal optimal board leadership structure, and companies should select the most appropriate model depending on their own organisational characteristics and business environment (see Finkeistein and D'Aveni, 1994; Boyd, 1995; Rhoades *et al.*, 2001; Palmon and Wald,

2002; Kiel and Nicholson, 2003; Faleye, 2004). Finkelstein and D'Aveni (1994) found that vigilant boards are more likely to adopt the CEO duality structure when informal CEO power and firm performance are low. They argue that under these conditions, the benefits of the unity of command from CEO duality outweigh the risks of CEO entrenchment. In contrast, board vigilance is negatively associated with CEO duality when both informal CEO power and firm performance are high.

Boyd (1995) reckoned that CEO duality is more advantageous under certain external environmental conditions, such as resource scarcity, and complex and dynamic environments, while Faleye (2004) found that companies with complex operations, sound CEO reputation, large managerial ownership and small board size are more likely to have CEO duality. Furthermore, Rhoades *et al.*, (2001) find that the duality-performance relationship is contingent on the decision context, and they detect that non-duality structure has a negative (positive) relationship with firm performance in anti-takeover (compensation) studies.

Similarly, Kiel and Nicholson (2003) provide empirical evidence that the board leadership structure of Australian companies is contingent on their company's size and business environment. Non-duality structure is more often adopted by large Australian companies, while CEO duality structure is more often adopted "in some highly entrepreneurial situations" (Kiel and Nicholson, 2003: 202). Likewise, Palmon and Wald (2002) also detect a positive relationship between small firm size and CEO duality, and conclude that large firms would benefit more from non-duality.

In family-controlled firms, controlling family members may appoint directors to the boards, and thus, the independence of directors may be compromised by their familiarity with family members (Chen and Jaggi, 2000; Leblanc and Gillies, 2005). Furthermore, family members have dominant control over the board and may hold top management positions, such as CEO and/or chairperson (La Porta *et al.*, 1999). As a result, CEO duality is more common in family-controlled firms than non-family firms (see also Cheung *et al.*, 2004; Lei and Song, 2004; Chen *et al.*, 2005).

The Board and Management

The independent directors should be briefed about the corporate developments regularly by either the Chairperson or CEO or member of management. A good practice is that the directors should have access to managers as well as the CEO. This would be a somewhat controversial proposition, particularly for CEOs who deal with the issue only when directors seek management's input on the CEO's performance. The full involvement of a broader management team should be positive if directors are sensitive to management's responsibilities. In fact, there are some boards who also invite the appropriate managers to speak about the current issues.

The managers from the key areas such as research, development, marketing and sales are also invited to strategic planning meetings, which are scheduled at least once a year. The intention is this would allow directors to become familiar with the managers and enable meaningful relationships to develop. Likewise, the relationship between the board and the chief financial officer (CFO) is one of the most important. As a director's independence increases, so does their distance from the business and from information provided by management. The key provider of that information is usually the CFO, who will have to respect the dual reporting lines to the CEO and directors. This delicate situation requires either a cohesive management team or a brave CFO. Many CEOs welcome the CFO's full and timely disclosure. Others believe that independent directors should be told only positive news so management appears to be in control and doing a great job. A CFO who fears a CEO fails to find a way to deliver to the board and ends up in disaster. The following framework gives the details of the board functions:

Figure 2.1

Functions of the Board

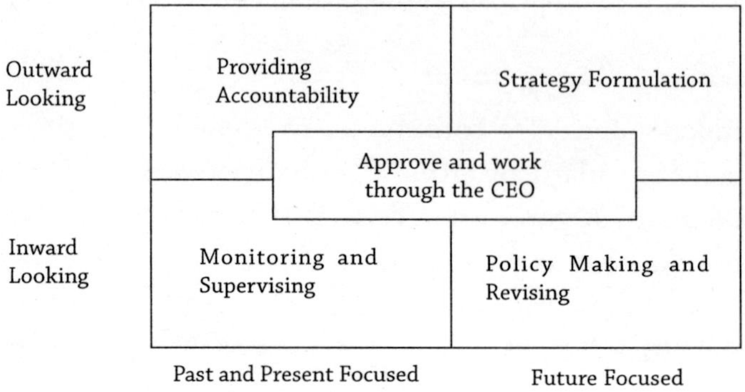

The separation of the functions of ownership and management has led to a growing role for board of directors. The board is becoming a body that acts in the interest of major shareholders and controls the activities of management. The boards of the companies with separated functions have proportionally twice as many outside and independent directors. Based on an Independent Directors Association (IDA) study completed in 2007, there are 256 independent directors in major 100 companies. Board committees are established in 85 per cent of the surveyed companies. The average board has 9 members, the range being from 4 to 16. The share of independent non-executive directors is about 28 per cent. Among independent directors in the larger companies, there is a trend towards increased representation of foreigners among independence non-executive director, from 40 per cent in 2006 to 52 per cent in 2007, which can be explained by IPOs (initial public offerings) recently held in foreign markets, mostly in London.

Experienced Western board practitioners who have worked with the top Russian executives and board members would observe the drive to improve governance in Russia. Forward-thinking companies are searching for ways of improving board's effectiveness and further improving their reputation. Directors have looked to improve their skill sets and companies have sought to improve the way their boards operate so that they can add more value. While a number of these companies need to meet standards set in countries in which they propose to list, there is a growing trend of leading Russian companies to get better value from their board. They also understand that high levels of governance need to add value far beyond the box-ticking exercise that one sees with codes of governance such as Sarbanes-Oxley and the UK's Combined Code.

The degree of board independence has also been linked to firm performance. Independence is not just a function of the proportion of inside *versus* outside directors, but also includes whether the board has dual leadership (CEO is also chairperson) and the degree of director's share ownership. Boards with dual leadership are considered less independent, as are boards with heavy share ownership. Anderson *et al.* (2004) found that independent audit committees had a lower cost of debt financing. In contrast, Mak and Roush (2000) found that boards with dual leadership, i.e. relatively less independent boards, were associated with more growth opportunities. The authors suggest that firms with greater agency problems, attributable to low inside share ownership and significant growth opportunities, are likely to choose boards that are more effective at mitigating the associated agency problems. Again, we see mixed results in terms of board of director independence and firm performance, especially in relation to dual leadership, leading one to question the utility of studying single board characteristics in relation to firm performance.

Theories Relevant to Duality Issue

Agency Theory

The development of agency theory is often traced back to Berle and Means (1932) although some writers suggest that one can go back to Adam Smith in 1776 and his influential book *The Wealth of Nations*.

Agency theory has a number of expressions. Previously classical economics conceived the firm as a single product entity with a commitment to the maximisation of profits, and what went on within the firm was considered to be of subordinate interest to what went on in markets. Financiers need the manager's specialised human capital to generate returns on their funds. In principle, the financiers and managers sign a contract that specifies what the managers do with the funds, and how the returns are divided between them and the financiers. The managers and financiers have to allocate residual control rights—the rights to make decisions not foreseen in the contract. Managers end up with substantial residual control rights, and therefore discretion over how to allocate investors funds. From this point of view, much of the subject of corporate governance concerns the constraints that managers put upon themselves, or that investors put on managers to reduce misallocation and thus to induce investors to provide more funds.

Agency theory offers shareholders a pre-eminent position in the firm legitimised not by the idea that they are the firm's owners, but instead its residual risk takers (Alchian and Demsetz, 1972). Fama (1980) emphasises the irrelevance of ownership:

> Ownership of capital should not be confused with ownership of the firm. Each factor in a firm is owned by somebody. The firm is just the set of contracts covering the way inputs are joined to create outputs and the way receipts from outputs are shared among inputs. In this 'nexus of contracts' perspective, ownership of the firm is an irrelevant concept.

The agency view suggests that shareholders are the 'principals' in whose interest the corporation should be run even though they rely on others for the actual running of the corporation. Fama and Jensen (1983) proposed the separation of decision-making and risk-bearing functions observed in the large corporation occurs in other organisations such as large professional partnerships, financial mutuals and non-profits:

> We contend that separation of decision-making and risk-bearing functions survives in these organisations in part because of the benefits of specialization of management and risk bearing but also because of an effective common approach to controlling the agency problems caused by separation of decision-making and risk-bearing functions.

It is claimed that shareholders have the right to residual claims because they are the residual risk bearers. As equity investors, shareholders are the only economic actors who make investments in the corporation without any contractual guarantee of a specific return, it is suggested. As 'residual claimants,' shareholders bear the risk of the corporation making a profit or loss, and have an interest in the allocation of corporate resources to make the largest residual possible. Since other stakeholders in the corporation will receive the returns for which they have contracted, the maximisation of shareholder value results in superior economic performance, not only for the particular corporation, but for the economy as a whole. Fama and Jensen (1983) argued that shareholders can diversify their investment portfolios to minimise risk. This separation of management and residual risk bearing in the corporation allows optimal risk allocation in the corporate economy.

Since, the basis of agency theory is the self-interested utility-maximising motivation of individual actors, it is assumed that the relationship between shareholders ('principals') and managers ('agents') will be problematic: how is the 'principal' able to prevent the 'agent' from maximising his own utility (Jensen, 1993)? For agency theorists, efficient markets are the solution; consequently the main focus of their approach to corporate governance is the elaboration and facilitation of market mechanisms that can mitigate the agency problem. These include an efficient market for corporate control, management labour and corporate information.

All of which will ensure that the management bears the costs of its own misconduct and will therefore create the incentives for self-

reassurance that managers will try to achieve outcomes that are in the shareholders' interests (Shleifer and Vishny, 1997). Internal and external governance mechanisms help to bring the interests of managers in line with those of shareholders, including:

- an effectively structured board;
- compensation contracts that encourage a shareholder orientation;
- concentrated ownership holdings that lead to active monitoring of executives;
- the market for corporate control that is an external mechanism activated when internal mechanisms for controlling managerial opportunism or failure have not worked (Daily *et al.*, 2003; Walsh and Seward, 1990).

As Learmount (2002) argues, Anglo-American economic theories of the firm have dominated thinking on corporate governance in much of the world, despite widespread criticisms of their underlying assumptions and propositions that they tend to see the firm principally in contractual terms; are guided by the assumption of utility-maximising self-interested human behaviour; tend to posit the protection of investor's capital as the 'corporate governance problem.'

However, whereas economic theories are well-established in academic research and public policy, organisational approaches to corporate governance tend to be less well-developed theoretically, or are relatively embryonic, and as a result have had much less influence on policy development. As Learmount contends, the impact of the hegemonic influence of exclusively economic approaches in corporate governance theorising has constricted the subject into a restricted form: "Notwithstanding their inherently distinctive conceptualizations of the 'governance problem', the different theories have helped to rectify the firm as an economic mechanism, and institutionalize the idea that corporate governance is primarily concerned with the control of managers by shareholders" (Learmount, 2002).

Stewardship Theory

Stewardship theory has its roots in psychology and sociology. It was designed for researchers to examine situations in which executives as stewards are motivated to act in the best interests of their principals (Donaldson and Davis, 1991). In stewardship theory, the model of man is based on a steward whose behaviour is ordered such that pro-organisational, collectivistic behaviours have higher utility than individualistic, self-serving behaviours. Given a choice between self-serving behaviour and pro-organisational behaviour, a steward's behaviour will not depart from the interests of his or her organisation. According to stewardship theory, the behaviour of the steward is collective, because the steward seeks to attain the objectives of the organisation (e.g., sales growth or profitability). This behaviour in turn will benefit principals such as outside owners (through positive effects of profits on dividends and share prices) and also principals who are managerial superordinates, because their objectives are furthered by the stewards. Stewardship theorists assume a strong relationship between the success of the organisation and the principal's satisfaction. A steward protects and maximises shareholders wealth through firm performance, because by doing so, the steward's utility functions are maximised (Davis *et al.*, 1997).

Given the potential multiplicity of shareholders' objectives, a steward's behaviour can be considered organisationally centred. Stewards in loosely coupled, heterogeneous organisations with competing stakeholders' and competing shareholders' objectives, are motivated to make decisions that they perceive are in the best interests of the group. Even in the most politically charged environment, one can assume that most parties desire a viable, successful enterprise. A steward who successfully improves the performance of the organisation generally satisfies most groups, because most stakeholder groups have interests that are well served by increasing organisational wealth. Therefore, a pro-organisational steward is motivated to maximise organisational performance, thereby satisfying the competing interests of shareholders. This does not imply that stewards do not have necessary 'survival' needs. Clearly, the steward must have an income to survive. The difference between the agent and the principal is how these needs are met. The steward realises the trade-off between personal needs and organisational objectives and believes that by working toward organisational, collective ends, personal needs are met. Hence, the steward's opportunity set is constrained by the perception that the utility gained from pro-organisational behaviour is higher than the utility that can be gained through individualistic, self-serving behaviour. Stewards believe their interests are aligned with that of the corporation and its owners. Thus, the steward's interest and utility motivations are directed to organisational rather than personal objectives.

Stewardship theorists argue that the performance of a stewardship is affected by whether the structural situation in which he or she is located facilitates effective action. If the executive's motivations fit the model of man underlying stewardship theory, empowering governance structures and mechanisms are appropriate. Thus, a steward's autonomy should be deliberately extended to maximise the benefits of a steward, because he or she can be trusted.

Given the advantage of stewardship to principals, why isn't there always a steward relationship, rather than agency relationship? The answer lies in the risks that the principals are willing to assume. In the governance contract between owners and executives, owners must decide how much risk they are willing to assume with their wealth. Risk-averse owners will most likely perceive that executives are self-serving and will prefer agency governance prescriptions. Implementing stewardship governance mechanisms for an agent would be analogous to turning the hen house over to the fox. Agency prescriptions can be viewed as the necessary costs of insuring principal utility against the risks of executive opportunism. From this perspective, a better question might be why would an owner ever take the risks of stewardship governance prescriptions?

Previously, empirical researchers have attempted to validate either agency theory or stewardship theory as a 'one best way' to corporate governance, assuming that all managers are either stewards or agents. The results of these studies have resulted in mixed findings; thus, there is the need for both agency theory and stewardship theory explanations of management (Donaldson and Davis, 1994). For example, several researchers found that the agency prescription of independent board leadership (i.e. a non-executive board chair) is associated with higher firm performance (e.g., Berg and Smith, 1978; Daily and Dalton, 1994; Rechner and Dalton, 1991). Other researchers found that stewardship's executive-chaired boards have significantly higher corporate performance (Donaldson and Davis, 1991; Finkelstein and D'Aveni, 1994). Still others suggest there is no significant difference in firm performance between executive- and outsider-chaired boards. Empirical evidence is similarly mixed with respect to other governance dimensions (Donaldson and Davis, 1994a). The mixed support for agency and stewardship theories suggests a need to reconcile these differences.

Factors that Differentiate between Agency and Stewardship Theories

There are a number of dimensions on which agency theory assumptions differ from the assumptions of stewardship theory and thereby serve to differentiate the theories. These dimensions can be characterised broadly as either psychological factors or situational factors and are presented below.

Psychological Factors

The fundamental difference between agency and stewardship theories with respect to psychological factors can be traced to the historical debates regarding the 'model of man'. According to the agency theory, man is rooted in economic rationality. In an interesting response to the work of Simon (1957a; 1957b), Argyris (1973a: 253) challenged this view of economic man as simplistic view of human behaviour and argued for a model of man, in order to increase the explanatory power and relevance of organisational theory. The model of man advocated by Argyris, characterised as "self actualizing man," has its roots in the early work of McGregor (1960) and the later work of Maslow (Willekens *et al.*, 2005). This model is based on the view that humans have a need to grow beyond their current state and reach higher levels of achievement and that the assumption of the economic view of man limits people from attaining their full potential. Argyris argued that when humans are placed in organisations that are designed on this economic view, they tend to suppress their level of aspirations to that of the economic view of man, and limit people from attaining their full potential. He further argued that for those individuals who are unable to suppress their aspirations and frustration with the organisational structures may lead to withdrawal and aggressive behaviours may be a problem.

Motivation: The major distinction between agency and stewardship theories is the focus on extrinsic *versus* intrinsic motivation. In agency theory, the focus is on extrinsic rewards—tangible, exchangeable commodities that have a measurable 'market' value. These extrinsic rewards form the basis for the reward systems that represent the control mechanisms of agency theory. For example, a principal may create a piecework incentive system to protect himself or herself from a self-serving agent. Similarly, medical insurance, 401k savings, and retirement plans may be instituted as a control mechanism to reduce the likelihood of turnover. Each of these rewards has a quantifiable value in terms of dollars that is recognised by both the parties. In contrast, in stewardship theory, the focus is on intrinsic rewards that are not easily quantified. These rewards include opportunities for growth, achievement, affiliation and self-actualisation. Subordinates in a stewardship relationship are reinforced by these intrinsic, intangible rewards and are motivated to work harder on behalf of the organisation. The basis for this distinction can be found in most of the established theories of motivation, but they are particularly apparent in the need theories. In a stewardship relationship, the focus would be on the higher order needs of Maslow's hierarchy (Willekens *et al.*, 2005), on Alderfer's growth need (1972), and on the achievement and affiliation needs of McClelland (1975).

In a more recent approach to the study of intrinsic motivation, Manz (1986; 1990) developed a theory of self-leadership. According to Manz, "self-leadership is a comprehensive self-influence perspective that concerns leading oneself towards performance of naturally motivating tasks as well as managing to do work that must be done but is not naturally motivating." Self-efficacy, self-determination and feelings of purpose are characterised as being critical determinants of intrinsic motivation. He argued that self-leadership involves a belief in one's work that extends beyond the formal reward system and relates to the importance of shared organisational vision. These views are consistent with the motivational assumptions of stewardship theory.

One other group of motivational theories that presents a unique perspective on the comparison between agency and stewardship assumptions is social comparison theories or equity theories (Adams, 1965; Cosier and Dalton, 1983). Although the basic premise of an exchange agreement that is a part of equity theory is more reminiscent of the economic view of man, the distinction between the perspectives is apparent in the social comparison that is assumed. In agency theory, there is an economic, or class-related separation between the principal and the agent. In developing an equitable work arrangement for the agent, the principal considers the fair market wage for the agent and arranges the compensation structure accordingly. The comparison for the agent in determining the 'fairness' of the situation is with respect to other agents in similar contexts. In stewardship theory, the principal is a part of the collective and the basis of comparison would include the principal. Thus, according to stewardship theory, the principal would expect to be accountable to the collective for his or her contributions as much as the steward may be qualitatively different and not easily quantifiable, the comparison and mutual accountability are expected.

The people who are motivated by higher order needs are more likely to become stewards in principal-steward relationships than people who are not motivated by higher order needs. People who are motivated by intrinsic factors are more likely to become stewards in principal-steward relationships than people who are motivated by extrinsic factors (Davis *et al.*, 1997).

Identification

Identification occurs when managers define themselves in terms of their membership in a particular organisation by accepting the organisation's mission, vision and objectives (Kelman, 1958), producing a satisfying relationship (O'Reilly, 1989). Through identification, an organisation becomes an extension of the steward's psychological structure (Brown, 1969). An identifying manager interprets comments about the organisation as referring also to himself or herself (i.e. he or she takes the comments personally). Identification allows managers vicariously to take credit for organisational failures (e.g., Katz and Kahn, 1978; Turner, 1981). Because managers vicariously take credit for organisational success, identification can increase the work-related satisfaction (e.g., Atkinson, 1957).

A manager who identifies with an organisation will thereby work toward the organisation's goals, solve its problems, and overcome barriers that are preventing the successful completion of tasks and assignments (Bass, 1960). When individuals identify with their organisations, they more readily engage in cooperative, altruistic and spontaneous un-rewarded citizenship behaviours (O'Reilly and Chatman, 1986). Therefore, managers who identify with their organisation are motivated to help it succeed and should be empowered to perform their jobs because this will enable them to use their initiative to promote the success of their organisations and their principles. A concept that is closely related to identification is organisational commitment. Porter *et al.* (1974) defined organisational commitment as the strength of the individual's identification with, and involvement in a particular organisation. Mayer and Schoorman (1992) characterised organisational commitment as a multidimensional construction consisting of continuous commitment, which represents the desire to remain in the organisation, and value commitment, "which is the belief in, and acceptance of, the goals of the organisation."

The people who have high identification with the organisation are more likely to become stewards in principal-steward relationships than people who have low identification with the organisation. People who are high in value commitment are more likely to become stewards in principal-steward relationships than people who are low in value commitment (Davis *et al.*, 1997).

Use of Power

Power is an important aspect of the relationship between a principal and a manager. A number of researchers have found that managers receive satisfaction from, and are motivated by, the use of power (e.g., McClelland, 1970; 1975; McClelland and Burnham, 1976). McClelland and Burnham (1976) defined the power motive as a psychological need to influence others toward the accomplishment of valid and accepted organisational goals. Managers who have a high need for power tend to "influence or direct other people; express opinions forcefully; enjoy the role of leader and may assume it spontaneously" (Steers and Black, 1994). The types of power used in the context of the relationship help to differentiate principal-agent relationships from principal-steward relationships. In the most widely cited typology of power bases, French and Raven (1959) described power in terms of coercive, legitimate, reward, expert and referent power. In a compatible but simpler typology, the five bases of power are reduced to institutional or organisational power and personal power (Gibson *et al.*, 1991). Institutional power is defined as being vested in the principal by virtue of his or her position in the organisation. Thus, termination of organisational membership would terminate the individual's power. The coercive, legitimate and many aspects of reward power described by French and Raven (1959) could be characterised as institutional power. In agency theory, institutional power is the basis of influence in the context of the principal-agent relationship. In this theory, reward power and legitimate power are used. Appropriate incentive systems and recognition of authority of the principal are combined to create the required level of control in the relationship. Coercive power represents the more severe method of agent control and is often present in a more subtle form through the threat of termination of employment. Personal power, an inherent part of the individual in the context of the interpersonal relationship, is not affected by position. Expert and referent power are characterised as personal power; referent power works through identification of one person with another person. Personal power is developed over time in the context of the relationship and is not affected by the formal roles in organisation. Although slower to develop, personal power can be sustained over longer periods of time. Personal power is the basis of influence in a principal-steward relationship. The choice of the type of power used is a function of the personal characteristics of the individual and the prevailing organisational culture. Certain organisational cultures facilitate the use of institutional power and, therefore, predispose members to principal-agent relationships.

People who are more likely to use personal power as a basis for influencing others are more likely to become stewards in principal-stewards relationships than people who use institutional power (Davis *et al.*, 1997).

Situational Factors

Management Philosophy

In the early debates between Argyris and the advocates of the economic model of man, one of the critical points of contention was whether organisational theory should be focused on descriptive or normative models of the organisation. Simon (1957; 1973) and others (e.g., Cyert and March, 1963) argued that the economic model and, therefore, implicitly the agency theory assumptions were the predominant basis of relationships in organisations. They cited numerous examples of behaviour by both principals and agents to support this claim. In contrast, Argyris (1973a) argued that the management philosophy of most organisations was based on economic assumptions and that this became a self-fulfilling prophecy regarding the nature of relationships that would develop. He advocated the developments of normative models of organisation based on self-actualising assumptions in order to create an organisational culture that supported the development of stewardship types of relationships. The position advocated by Argyris (1973a) was similar to the arguments advanced earlier by McGregor (1960) in his discussion of Theory Y management and by Likert (1961) in his comparison of system management with more control-oriented systems. The common message in each of these theories is that the assumption about the model of man drives the development of management philosophies and management systems, which then serve to produce behaviour in the organisation that is consistent with the assumptions. Each of these theories advocated the development of normative models of organisation and a break from the traditional management philosophies in order to facilitate the self-actualising behaviours that are consistent with stewardship theory.

More recently, Walton (1980; 1985) advocated what he called a high commitment management philosophy. This approach to management was characterised as being highly participative and consisting of open communication, empowerment of workers and the establishment of trust. Lawler (1986; 1992) elaborated on this view by contrasting the management philosophies he described as control-oriented *versus* involvement-oriented. According to Lawler, the control-oriented philosophy is based on a management philosophy that the thinking and controlling part of the work must be separated from doing part of the work. The key assumption in involvement-oriented approaches is that when employees are given challenges and responsibility, they will develop self-control and monitor their behaviour.

Table 2.1

Comparison of Agency Theory and Stewardship Theory

S.No.	*Criteria*	*Agency Theory*	*Stewardship Theory*
1	Model of Man	Economic Man	Self-Actualising Man
2	Behaviour	Self-serving	Collective Serving
3	Motivation	Lower Order/Economic Needs (Psychological, Security, Economic)	Higher Order Needs (Growth Achievement, Self-actualisation)
4	Social Comparison	Other Managers	Principal
5	Identification	Low Value Commitment	High Value Commitment
6	Power	Institutional (Legitimate, Coercive, Reward)	Personal (Expert, Referent)
7	Management Philosophy	Control Oriented	Involvement Oriented
8	Risk Oriented	Control Mechanisms	Trust
9	Timeframe	Short Term	Long-term
10	Objective	Cost Control	Performance Enhancement
11	Cultural Difference	Individualism High Power Distance	Collectivism Low Power Distance

Source: Duppati and Mishra (2008), compiled from literature.

CASE POINT

Review of Issues in Corporate Governance: India and New Zealand

This section reviews the recently published works on India and New Zealand corporate governance practices (Scrimgeour and Duppati, 2014).

Corporate governance is obviously a matter of global concern and has gained tremendous importance in recent years in the context of globalisation of economies and financial markets. The financial crisis of 2008 and the 2012 European crisis involving Greece, Italy and Spain revealed corporate governance failures in financial institutions and corporations, leading to systemic consequences (Classens and Yurtoglu, 2013). Earlier, two major scandals, Enron and WorldCom in the US resulted in the enactment of Sarbanes-Oxley Act, 2002, as a measure to ensure and restore investors' confidence in business in particular and the interest of society at large. This lead to corporate governance reforms worldwide impacting corporate board composition, conduct, and responsibility at the legal and regulatory levels (Frank and Duppati, 2014).

Majority of the studies in corporate governance are centred around the principal-agent relationships and principal-principal agency relationships. But the consideration given to the institutional and contextual framework is minimal (Globerman *et al.*, 2011; JAguilera *et al.*, 2008). Likewise, the economic, social and political conditions vary by country, and a more subtle understanding of how these factors are continuously shaping the business environment is critical to spotting new opportunities and managing unexpected risks.

The study of Scrimgeour and Duppati (2014) has taken a wider approach and has seen if the corporate governance framework differs in different settings. Their paper is built up on the wider perspective of the principal-agent relationships involving four layers of these relationships in state-owned enterprises (SOEs). The idea is straight forward. First layer involves general public (who are deemed to be the real owners of the SOEs) and government, second layer involves government and ministers, third layer involves ministers and boards, and fourth layer involves boards and the management. Therefore, the governments and ministers who function as principals to the managers in SOEs are themselves part of a more evolved chain of delegation which runs from voters and their representatives in parliament to the cabinet and individual ministers and further to the administrative apparatus of the state and then the boards and management (Huber, 2000; Lupia, 2003; Müller, 2000; Strøm, 2000; Mishra and Duppati, 2007).

Figure 2.2

Theoretical Model: Four Layers of Principal Agent Relationships

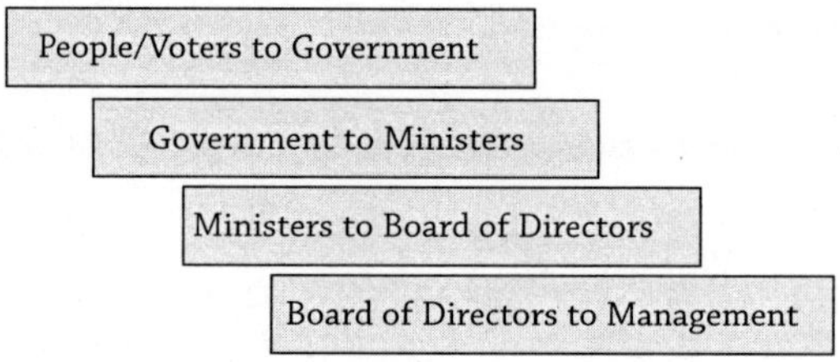

Source: Scrimgeour and Duppati (2014).

In spite of the corporate governance reforms, the underperformance of SOEs continued to be a prominent concern in the available literature. A state-owned firm faces organisational costs associated with two types of internal conflict of interest, namely, political costs associated with government (owner) incentive to intervene in the firm and agency costs associated with a manager's incentive to expropriate wealth from the firm (Shleifer and Vishny, 1994; Fan, 2012). Whether these costs can be contained is the key to the success of SOE reform. Jedenastik (2013) argues that managers in public corporations are part of the chain of delegation that structures much of the political process and his results support the proposition that partisan affliation drives managerial survival. The study undertaken by Reddy *et al.* (2011) investigated the nature of corporate governance practices in public sector corporate entities in New Zealand and their effects on financial performance.

Using agency theory as the dominant theoretical paradigm, the extant literature has mainly focused on the efficacy of various governance mechanisms that protect the shareholders from self-serving managers (Rajagopalan and Zhang, 2008). Much of the research is situated in the context of developed economies, where the external governance environment and institutions to support the internal firm governance are stable and well developed (Judge *et al.*, 2008).

While a focus on within–firm governance mechanisms has advanced our understanding of the links between governance standards and firm performance, there is an increasing realisation that the efficacy of within firm governance may be dependent on the quality of external governance and institutions (Judge *et al.*, 2008). This issue is particularly important for emerging economies, which often lack the institutions needed to support efficient within firm governance (Peng, 2003). It is well documented that many emerging economies, such as India and China, do not have well developed external control mechanisms, such as a market for corporate control, merger and acquisition laws, and efficient law enforcement (Khanna and Palepu, 2000a; Peng, 2003).

This not only makes it more difficult to govern the organisations, but also makes standard corporate governance practices less legitimate (Judge *et al.*, 2008). It is evident from the literature presented above that there is gap in the literature from cross-country comparative perspective involving SOEs. Majority of the studies involving agency perspective are undertaken from the context of listed companies.

This study extends the literature beyond the firm level agency perspective. It considers the Organisation for Economic Co-operation Development (OECD) guidelines on corporate governance of SOEs as a basic framework, and explores how the challenges of corporate governance have been addressed in state-owned corporations of India and New Zealand. The study seeks to accurately describe the different approaches taken and report on their perceived effectiveness.

Conclusion

The study identifies that the New Zealand state-owned entities are categorised based on their objectives, thereby differentiating between the commercial and social objectives. The ownership policy of New Zealand is explained in the Companies Act and SOE Act of 1986. Infact, the clearly articulated ownership policies provides a background for prioritising SOEs' objectives and are instrumental in limiting the consequences of passive ownership or excessive intervention in SOEs' management. Likewise, the

conduct of the candidate has to be furnished to the cabinet committee on appointment. If all this goes smoothly, it takes about 8-12 months before the appointment takes place. This is indicative of corporate governance issues in India.

Recently, Dr U.D. Choubey, Director General, SCOPE, has expressed his views[1] on independent directors role in Indian SOEs. Independent directors are key ingredients of corporate governance at the board level. They are supposed to be watchdogs or conscience-keepers sitting at the fence with no accountability and pressure from either the administrative ministry or functional directors. But in practice, their loyalty lies with the administrative ministry so much so that we can brand them as 'dependable independent directors.' The fact is that a lot of lobbying goes into their appointment process, which results in a situation where there is a big gap between them and the functional area of an enterprise. The result is that they come to the board meetings unprepared and take interest in only agenda items which suit the interests of their appointing authority. This is a lacuna in the present system and he felt that the root cause is the selection system.

In order to avoid the agency costs arising due to political intervention, it is suggested that the power to select independent directors to the boards should be completely vested in the PESB. To avoid any conflict of interest, the administrative ministry should not be in the search committee of the PESB nor be a votary to the selection of a particular candidate. The PESB should be given the powers of a constitutional body like their Appointments Committee of the Cabinet (ACC) so that its selection of an independent director is final. Since this process may need parliamentary approval, we could look at a easier option where the PESB selects a candidate and sends it to the cabinet secretary for notification, avoiding the administrative ministry.

Therefore, the issues identified in the literature of SOEs which includes bureaucracy, political interference and political patronage continue to persist in India in spite of the corporate governance reforms. It is evident that the agency conflicts arising from four layers are evident in case

1 "Delink Independent Directors from ministry", *Financial Express*, 4 December 2013, Wednesday.

of India, while New Zealand case shows evidence of transparency and preferred settings for a better accountability and fits into the frame of OECD guidelines.

References

Adams, J.S. (1965). "Injustices in Social Exchange", in L. Berkowitz (ed.), *Advances in Experimental Social Psychology*. New York: Academic Press. pp.267-99.

Aguilera, R.A., I. Filatotchev, H. Gospel and G. Jackson (2008) "An Organizational Approach to Comparative Corporate Governance: Costs, Contingencies, and Complementarities", *Organization Science* 19: 475–92.

Alchian, A.A. (1968). "Corporate Management and Property Rights", in *Economic Policy and the Regulation of Securities*. Washington, DC: American Enterprise Institute.

Alchian, Armen A. and Harold Demsetz (1972). "Production, Information Costs and Economic Organisation", *American Economic Review* 62: 777-95.

Alderfer, Clayton P. (1972). *Human Needs in Organizational Settings*. New York: The Free Press of Glencoe.

Anderson, R.C., S.A. Mansi and D.M. Reeb (2004). "Board Characteristics, Accounting Report Integrity and the Cost of Debt", *Journal of Accounting and Economics* 37: 315-42.

Ang, J.S., R.A. Cole and J. Wuh Lin (2000). "Agency Costs and Ownership Structure", *Journal of Finance* 55: 81-106.

Argyris, C. (1973a). "Organization Man: Rational and Self-actualizing", *Public Administration Review* 33 (July/August): 354-57.

———. (1973b). "Some Limits of Rational Man Organisational Theory", *Public Administration Review* 33 (May/June): 253-67.

Atkinson, J.W. (1957). "Motivational Determinants of Risk-taking Behaviour", *Psychological Review* 64: 359-72.

Baliga, B.R., R.C. Moyer and R.S. Rao (1996). "CEO Duality and Firm Performance: What's the Fuss?", *Strategic Management Journal* 17(1): 41-53.

Bass, B.M. (1960). *Leadership, Psychology and Organisational Behaviour.* New York: Harper.

Berg, S.V. and S.K. Smith (1978). "CEO and Board Chairman: A Quantitative Study of Dual vs. Unitary Board Leadership", *Directors & Boards* 3(l):34-39.

Berle, A.A. and G.C. Means (1932). *The Modern Corporation and Private Property.* New York: Macmillan.

Boyd, B.K. (1995), "CEO Duality and Firm Performance: A Contingency Model", *Strategic Management Journal* 16(4): 301-12.

Brown, M.E. (1969). "Identification and Some Conditions of Organizational Involvement", *Administrative Science Quarterly* 14: 346-55.

Cadbury, A. (1992). *Report of the Committee on the Financial Aspects of Corporate Governance: The Code of Best Practice (Cadbury Code)*. London: Gee & Company.

Chen, C.J.P. and B. Jaggi (2000), "Association Between Independent Non-executive Directors, Family Control and Financial Disclosures in Hong Kong", *Journal of Accounting and Public Policy* 19(4-5): 285-310.

Cheung, S.Y.L., J.T. Connelly, P. Limpaphayom and L. Zhou (2004). *Do Investors Really Care about Corporate Governance? Evidence from the Hong Kong Market*. Hong Kong: City University of Hong Kong and The Hong Kong Institute of Directors.

Coase, R.H. (1937). "The Nature of the Firm", *Economica* 4(16): 386-405 (New Series). Reprinted in George J. Stigler and Kenneth E. Boulding (eds.) *Readings in Price Theory*. Chicago, Homewood, IL: Richard D. Irwin Inc. pp.331-351.

Claessens, S. and B. Yurtoglu (2012). *Corporate Governance and Development : An Update*. A Global Corporate Governance Forum Publication, IFC.

Cosier, R. A. and D.R. Dalton (1983). "Equity Theory and Time: A Reformulation", *Academy of Management Review* 8: 311-19.

Cyert, R. M. and J.G. March (1963). A *Behavioral Theory of the Firm*. Englewood Cliffs, NJ: Prentice Hall.

Daily, C.M. and D.R. Dalton (1992). "The Relationship Between Governance Structure and Corporate Performance in Entrepreneurial Firms", *Journal of Business Venturing* 7(5): 375-86.

———. (1994a). "Bankruptcy and Corporate Governance: The Impact of Board Composition and Structure", *Academy of Management Journal* 37(6): 1603-17.

———. (1994b). "Corporate Governance in the Small Firm: Prescriptions for CEOs and Directors", *Journal of Small Business Strategy* 5(1): 57-68.

Dalton, D.R., C.M. Daily, J.L. Johnson and A.E. Ellstrand (1999). "Number of Directors and Financial Performance: A Meta Analysis", *Academy of Management Journal* 42(6): 674-86.

Daily, C. M., D.R. Dalton and A.A. Cannella (2003). "Corporate Governance: Decades of Dialogue and Data", *Academy of Management Review* 28: 371–82.

Donaldson, L. (1990). "The Ethereal Hand: Organizational Economics and Management", *Theory, Academy of Management Review* 15: 369–81.

Donaldson, L. and J. H. Davis (1991). "Stewardship Theory or Agency Theory: CEO Governance and Shareholder Returns", *Australian Journal of Management* 16: 49–64.

———. (1994). "Boards and Company Performance: Research Challenges the Conventional Wisdom", *Corporate Governance: An International Review* 2: 151–60.

Dulewicz, V. and P. Herbert (2004). "Does the Composition and Practice of Boards of Directors Bear any Relationship to the Performance of their Companies?", *Corporate Governance: An International Review* 12(3): 263-80.

Faleye, 0. (2004). "Does One Hat Fit All? The Case of Corporate Leadership Structure", *Working Paper.* Boston, MA: College of Business Administration, Northeastern University.

Fama, E.F. (1980). "Agency Problems and Theory of the Firm", *Journal of Political Economy* 88(2): 288-307.

Fama, E. and M. Jensen (1983). "Separation of Ownership and Control", *Journal of Law & Economics 26:* 301-25 and 327-49.

Finkelstein, S. and R.A. D'Aveni (1994). "CEO Duality as a Double-edged Sword: How Boards of Directors Balance Entrenchment Avoidance and Unity of Command", *Academy of Management Journal* 37(5): 1079-08.

Frederick, W. (2011). "Enhancing the Role of the Boards of Directors of State-Owned Enterprises", *OECD Corporate Governance Working Paper* No.2. OECD Publishing.

French, J.R.P. Jr. and B.H. Raven (1959). "The Bases of Social Power", in D. Cartwright (ed.), Studies in Social Power. Ann Arbor, MI: Institute for Social Research. pp.150–67.

Gibson, J. L., J.M. Ivancevich and J.H. Donnelly (1991). *Organizations*. Homewood, IL: Irwin.

Globerman, S., M.W. Peng, and D. , Shapiro (2011). "Corporate Governance and Asian Companies", *Asia Pacific Journal of Management* 28(1): 1-14.

Gul, F. A. and S. Leung (2004). "Board Leadership, Outside Directors' Expertise and Voluntary Corporate Disclosures", *Journal of Accounting and Public Policy* 23: 351-79.

Higgs, D. (2003). *Review of the Role and Effectiveness of Non-executive Directors*. London: The Department of Trade and Industry.

Huber, J. (2000). "Delegation to Civil Servants in Parliamentary Democracies", *European Journal of Political Research* 37(3): 397–13.

Jedenastik, L.E. (2013). "Political Control and Managerial Survival in State-Owned Enterprises", *Governance: An International Journal of Policy, Administration, and Institutions* 27(1). Wiley Periodicals Inc. doic: 10.1111/gove.12023.

Jensen, M. (1993). "The Modern Industrial Revolution: Exit and the Failure of Internal Control Systems", *Journal of Finance* 3: 831–80.

Judge, W.Q., T.J. Douglas and A.M. Kutan (2008). "Institutional Antecedents of Corporate Governance Legitimacy", *Journal of Management* 34: 765–85.

Kang, E. and A. Zardkoohi (2005). "Board Leadership Structure and Firm Performance", *Corporate Governance: An International Review* 13(6): 785-99.

Katz, D. and R.L. Kahn (1978). *The Social Psychology of Organizations*. 2nd ed. New York: Wiley.

Kelman, H.C. (1958). "Compliance, Identification, and Internalization: Three Processes of Attitude Change, *Journal of Conflict Resolution* 2, 51-60.

Khanna, T. and K. Palepu (2000). "Is Group Affiliation Profitable in Emerging Markets? An Analysis of Diversified Indian Business Groups," *Journal of Finance* 55: 867-91.

Kiel, G.C. and G. Nicholson (2003). "Board Composition and Corporate Performance: How the Australian Experience Informs Contrasting Theories of Corporate Governance", *Corporate Governance: An International Review* 11(3): 189-205.

La Porta, R., F. Lopez-De-Silanes and A. Shleifer (1999). "Corporation Ownership Around the World", *Journal of Finance* 54(2): 471-517.

Lawler, E. E. (1986). *High Involvement Management*. San Francisco: Jossey-Bass.

————. (1992). *The Ultimate Advantage*. San Francisco: Jossey-Bass.

Leblanc, P. and J. Gillies (2005). *Inside the Boardroom: How Boards Really Work and the Coming Revolution in Corporate Governance*. Montreal: John Wiley & Sons.

Learmount, S. (2002). "Theorizing Corporate Governance: New Organizational Alternatives", *Working Paper* 237. Cambridge: ESRC Centre for Business Research, Judge Institute for Management, University of Cambridge.

Lei, A.C.H. and F.M. Song (2004). "Corporate Governance and Firm Valuations: Evidence from Hong Kong", *Working Paper*. Hong Kong: School of Economics and Finance, The University of Hong Kong.

Likert, R. (1961). *New Patterns of Management*. New York: McGraw-Hill.

Lupia, A. (2003). "Delegation and its Perils", in Kaare Strøm, Wolfgang C. Müller and Torbjörn Bergman (eds.), *Delegation and Accountability in Parliamentary Democracies*. Oxford University Press.

Mak, Y.T. and M.L. Roush (2000). "Factors Affecting the Characteristics of Boards of Directors: An Empirical Study of New Zealand Initial Public Offering Firms", *Journal of Business Research* 47(2): 147-59.

Manz, C.C. (1986). "Self-leadership: Toward an Expanded Theory of Self-influence Processes in Organizations", *Academy of Management Review* 11: 585-600.

Manz, C.C. and H.P. Sims Jr. (1990). *SuperLeadership*. New York: Berkeley Books.

Mayer, R.C. and F.D. Schoorman (1992). "Predicting Participation and Production Outcomes through a Two-dimensional Model of Organizational Commitment", *Academy of Management Journal* 35: 671-84.

McClelland, D.C. (1975). *Power: The Inner Experience*. New York: Irvington-Halsted-Wiley.

McClelland, D.C. and D.H. Burnham (1976). "Power is the Great Motivator", *Harvard Business Review* 54(2): 100-10.

McGregor, D. (1960) *The Human Side of Enterprise*. New York: McGraw-Hill.

McKnight, P. J. and C. Weir (2009). "Agency Costs, Corporate Governance Mechanisms and Ownership Structure in Large UK Publicly Quoted Companies: A Panel Data Analysis", *The Quarterly Review of Economics and Finance* 49: 139-58.

Mishra, R.K. and Geeta Duppati (2007). "Implications of the Property Rights Theory in the Post-privatization Period: A Case of the Selected Enterprises in the State of Andhra Pradesh", *The Journal of Institute of Public Enterprise* 30(1&2): 72-85.

Mishra, R.K., S. Locke and Geeta Duppati (eds.) (2012). *Corporate Governance: A Global Perspective*. New Delhi, India: Academic Foundation.

Mishra, R., S. Sarkar and P. Singh (2012). "Linking CSR to Business Strategy: The Balanced Scorecard Approach", in *Strategic Corporate Social Responsibility: Towards a Sustainable Business*. Hyderabad: Bloomsbury Publishing India Pvt. Ltd. pp. 3-16.

Muller, W.C. (2000). "Political Parties in Parliamentary Democracies: Making Delegation and Accountability Work", *European Journal of Political Research* 37(3): 309-33.

O'Reilly, C. (1989). "Corporations, Culture and Commitment: Motivation and Social Control in Organizations", *California Management Review* 31(4): 9-25.

O'Reilly, C. and J. Chatman (1986). "Organizational Commitment and Psychological Attachment: The Effects of Compliance, Identification, and Intemalization on Prosocial Behavior", *Journal of Applied Psychology* 71: 492-99.

Palmon, O. and J.K. Wald (2002). "Are Two Heads Better Than One? The Impact of Changes in Management Structure on Performance by Firm Size", *Journal of Corporate Finance* 8(3): 213-26.

Peng, M. W. (2004). "Outside Directors and Firm Performance During Institutional Transitions", *Strategic Journal of Management* 25:453–71.

Porter, L., R. Steers, R. Mowday and P. Boulian (1974). "Organisational Commitment, Job Satisfaction, and Turnover among Psychiatric Technicians", *Journal of Applied Psychology* 59, 603-09.

PricewaterhouseCoopers LLP (2005). *Corporate Governance at the Crossroads : A Report*. Antwerpen Holmes Beach, Fl.: Intersentia.Scrimgeour, F. and Geeta Duppati (2014). "Corporate Governance in the Public Sector: Dimensions, Guidelines and Practice in India and New Zealand", *Corporate Ownership & Control* 11(2): 364-77.

Rajagopalan, N. and Y. Zhang (2008). "Corporate Governance Reforms in China and India: Challenges and Opportunities", *Business Horizons* 51: 55–64.

Rechner, P.L. and D.R. Dalton (1991). "CEO Duality and Organizational Performance: A Longitudinal Study", *Strategic Management Journal* 12: 155–60.

Reddy, K., S. Locke and F. Scrimgeour (2011). "Improving Performance in New Zealand's Public Corporations: The Effect of Governance Practices", *Governance: An International Journal of Policy, Administration, and Institutions* 24(3) 517-56.

Rhoades, D.L., P.L. Rechner and C. Sundaramurthy (2001), "A Meta-analysis of Board Leadership Structure and Financial Performance: Are 'Two Heads Better Than One'?", *Corporate Governance: An International Review* 9(4): 311-19.

Scrimgeour, F. and Geeta Duppati. (2014). "Corporate Governance in the Public Sector: Dimensions; Guidelines and Practice in India and New Zealand", *Corporate Ownership & Control* 11(2): 364-77.

Shleifer, A. and R. Vishny (1994). "Politicians and Firms", *Quarterly Journal of Economics* 109: 995-1025.

———. (1997). "A Survey of Corporate Governance", *The Journal of Finance* 52(2): 737-83.

Simon, H. A. (1957a). *Administrative Behavior*. 2nd ed. Glencoe, IL: Free Press.

———. (1957b). *Models of Man*. New York: Wiley.

———. (1973). "Organization Man: Rational or Self-actualizing?", *Public Administration Review* 33 (July/August): 346-53.

Steers, R.M. and J.S. Black (1994). *Organizational Behavior*. New York: HarperCollins.

Strøm, K. (2000). "Delegation and Accountability in Parliamentary Democracies", *European Journal of Political Research* 37(3): 261–290.

Trivedi, P. and M.P. Vittal (1992). "Menu of Financial Indicators used in MOUs: An Exercise in Clarification", *Economic and Political Weekly* 27(22): M59-M62.

Turner, J.C. (1981). "The Experimental Social Psychology of Intergroup Behavior", in J. C. Turner and H. Giles (eds.), *Intergroup Behaviour.* Chicago: University of Chicago Press. pp. 66-101.

Walsh, J.P. and J.K. Seward (1990). "On the Efficiency of Internal and External Corporate Control Mechanisms", *Academy of Management Review* 15: 421–58.

Walton, R. E. (1980). "Establishing and Maintaining High-commitment Work Systems", in J.R. Kimberly, R.H. Miles & Associates (eds.), *The Organizational Life Cycle: Issues in the Creation, Transformation, and Decline of Organizations.* San Francisco: Jossey- Bass. pp.208-90.

———. (1985). "From Control to Commitment in the Workplace", *Harvard Business Review* 63(2): 76-84.

Weisbach, M. (1988). "Outside Directors and CEO Turnover", *Journal of Financial Economics* 20(1/2): 431-60.

Williamson, O.E. (1975). *Markets and Hierarchies: Analysis and Antitrust Implications.* New York: Free Press.

———. (1979). "Transaction-Cost Economics: The Governance of Contractual Relations", *Journal of Law and Economics* 22: 233-61.

———. (1984). "Corporate Governance", *Yale Law Review* 93: 1997-230.

———. (1985). *The Economic Institutions of Capitalism*. New York: Free Press.

Willekens, M., P. Sercu, K. Desender and Katholieke Universiteit te Leuven (1970). *Faculteit der Economische en Toegepaste Economische Wetenschappen*.

3

Corporate Governance

Committees and Reports

Introduction

This chapter presents an overview of some of the codes and regulations designed to improve corporate governance in the UK, US and India. It reviews the recommendations of the various committees that were formed to intensify the practices of corporate governance. In this process, it presents and reviews the Cadbury Report, the Greenbury Report, the Hampel Committee Report, the Turnbull Report, the Higgs Report, the Smith Report and the Sarbanes-Oxley Act of 2002. It also makes a review of the Indian committees.

The Cadbury Report 1992

The Cadbury Committee was the first committee to be constituted to report on the financial aspects of corporate governance. It was set up in 1991 under the chairmanship of Sir Adrian Cadbury. The Cadbury Committee, which reported in 1992, acknowledged that the financial scandals and abuse of power as exposed in the Maxwell case were some of the reasons behind the setting up of the committee to report on corporate governance matters. Hence, the formation of the Cadbury Committee can be seen as reactive rather than proactive. However, it is important to remember that Cadbury Report was compiled on the basic assumption that the existing, implicit system of corporate governance in the UK was sound and that many of the recommendations were merely making explicit a good implicit system (see Cadbury Report, 1992: 12, para 1.7). The Cadbury Report and its accompanying Cadbury Code (1992) derived their names from Sir Adrian Cadbury. The council of the Stock Exchange and the Accountancy Profession set about establishing the Cadbury Committee, The Committee on the Financial Aspects of Corporate Governance, which produced its report, and accompanying Code of Best Practice at the end of 1992. The Cadbury Code was not legally binding on boards of directors. Nevertheless, one of the rules in the Stock Exchange Yellow Book (refer notes) at the time of its publication was a statement of compliance with the code. The result of this was that all companies publicly quoted on the Stock Exchange had to state in their annual reports whether or not they had implemented the code in all respects. If they had not complied with the whole code, then they were compelled to make a clear statement of the reasons while detailing and explaining the points of non-compliance. The implication was that the companies' shareholders then had the opportunity of deciding whether or not they were satisfied with the companies' corporate governance systems.

The Cadbury Report and its accompanying code covered three general areas, namely: the board of directors, auditing, and shareholders. The Cadbury Report focused attention on the board of directors as being the most important corporate governance mechanisms, requiring constant monitoring and assessment. However, the accounting and auditing functions were also shown to play an essential role in good corporate governance, emphasising the importance of the corporate transparency and communication with shareholders and other stakeholders. Lastly, Cadbury's focus on the importance of the institutional investors as the largest and most influential group of shareholders has had a lasting impact. This, more than any other initiative in corporate governance reform, has led to the shift of directors' dialogue toward greater accountability and engagement with shareholders. Furthermore, we believe that this move towards greater shareholder engagement has encouraged a more significant metamorphosis of corporate responsibilities toward a range of stakeholders, encouraging greater corporate social responsibility

(CSR) in general. There is no denying about the substantial impact that the Cadbury Code has had on corporate Britain and, indeed, on companies around the world. By the late 1990s, there was strong evidence to show a high level of compliance with the Cadbury Code's recommendations (see Conyon and Mallin, 1997).

Central to the final report's recommendations was that boards of all listed companies registered in the UK should comply with the Code of Best Practice as set out in the report. The code is given added weight by the disclosure requirement of the London Stock Exchange that companies must state in their annual report whether they are complying with the code and to give reasons for any aspects on non-compliance.

Code of Best Practice

At the time of publication of the Committee's final report, Sir Adrian Cadbury said:

> The planks on which the code is based are the need for disclosure and for checks and balances. Disclosure ensures that all those with a legitimate interest in a company have the information they need in order to exercise their rights and responsibilities toward it. In addition, openness by companies is the basis of public confidence in the corporate system. Checks and balances guard against undue concentrations of power and make certain that all the interests which boards have a duty to consider are properly taken into account.

The code recommendations consist of 19 points set out under the headings of: (1) The Board of Directors; (2) Non-Executive Directors (NEDs); (3) Executive Directors; and (4) Reporting and Controls. The main points are summarised as follows:

The Board of Directors

- The board should meet regularly, retain full and effective control over the company and monitor the executive management.

- There should be a clearly accepted division of responsibilities at the head of the company, which will ensure a balance of power and authority, such that no one individual has unfettered powers of decision.' Ideally the roles of chairman and chief executive should be separated, although this may not always be practical, in which case there 'should be a strong and independent element on the board.'
- The board should include NEDs of sufficient calibre and number for their views to carry significant weight in the board's decisions.

NEDs

- NEDs or 'outside' directors, as the committee's chairman preferred to call them, should 'bring an independent judgment to bear on issues of strategy, performance, resources, including key appointments and standards of conduct.'
- The majority of NEDs should be 'independent' of management and free from any business or other relationship which could materially interfere with the exercise of their independent judgment, apart from their fees and shareholding.
- NEDs should be appointed by a formal process and their appointment should be a matter for the board as whole. Appointments should be for specified terms and re-appointment should not be automatic.

Executive Directors

- Directors' service contracts should not exceed three years without shareholders' approval.
- Directors' pay and emoluments, including pension contributions, stock options, and the amount, as well as the basis for any performance-related element, should be fully disclosed and subject to the recommendations of remuneration committees consisting mainly or wholly of NEDs directors and preferably chaired by a NED.

Reporting and Controls

- It is the board's duty to present a balanced and understandable assessment of the company's position.
- The board should ensure that an objective and professional relationship is maintained with the auditors.
- The board should establish an audit committee which should consist of at least three NEDs. Originally, the committee referred to the annual audit as 'one of the cornerstones of corporate governance.'
- The directors should report on the effectiveness of the company's system of internal control.
- The directors should report that the business is a going concern, with supporting assumption or qualifications necessary.

The Greenbury Report 1995

A particularly contentious aspect of corporate governance in recent years has been that of executive pay. In 1994-95, the seemingly endless escalation in executive pays, particularly in the newly privatised public utilities such as British Gas, caused a public outcry in the UK. It forced the British Prime Minister at the time, John Major, to denounce 'un-justifiable' increases in company executive pay in the House of Commons in November 1994. In response to such public concern, the Confederation of British Industry (CBI) recruited a committee of 11 top managers (mainly chairmen) from leading UK companies such as British Petroleum, British Telecommunications, GKN, Boots, and Marks & Spencer to conduct an inquiry into directors' remuneration. The committee was chaired by Sir Richard Greenbury, the Executive Chairman of Marks & Spencer, and became known as the Greenbury Committee. Its brief was: 'To identify good practice in determining Directors' remuneration and prepare a code of such practice for use by UK public limited companies (PLCs).'

The committee published its report on 17 July 1995, and its key themes were: 'accountability, responsibility, full disclosure, alignment of Director and shareholder interests, and improved company performance' (Directors' Remuneration Greenbury, 1995). The Greenbury Committee was formed after widespread public concern over what was seen as excessive amounts of remuneration paid to directors of quoted companies and newly privatised companies. "Recent concerns about executive remuneration have centered, above all, on some large pay increases and large gains from share options in the recently privatized utility industries. These increases have sometimes coincided with staff reductions, pay restraints for other staff and price increases... there have also been concerns about the amounts of compensation paid to some departing directors' (Greenbury Report, 1995: 9). The Greenbury Committee was keen to ensure that directors' remuneration was linked to company performance and the committee did not seem to see a problem with high levels of pay per se, as long as they could be justified on the basis of the company's financial results.

A key concern should be to ensure, through the remuneration system, that directors share the interest of shareholders in making the company successful. Performance-related remuneration can be highly effective in aligning interest in this way. In many companies therefore, there will be a case for a high gearing of performance related to fixed pay. But there are two constraints on this. First, there will usually be a level of basic salary below which it will not be practicable to go. Second, the requirements and priorities of companies vary. The gearing, which suits one company, may be quite unsuitable for another (Greenbury Report, 1995: 38) The Greenbury Report also addressed the problem of departing directors whose performance had not been noticeably successful, but who still manage to leave the company with generous compensation for loss of office; compensation payments to directors on loss of office have been a cause of public and shareholders concern in recent times. Criticism has been directed at the scale of some of the payments made and at their apparent lack of justification in terms of performance. Some payments have been described as "rewards for failure" (Greenbury Report, 1995: 45). When the Greenbury Report was published in 1995, it dealt specifically with the question of directors' remuneration, and many of its recommendations were developed from the earlier Cadbury Report. The Greenbury Report recommended that the remuneration committee should consist exclusively of NEDs (the Cadbury Report had recommended wholly or mainly NEDs). These NEDs should have no personal financial interest, no potential conflicts of interest arising from cross-directorships and no day-to-day involvement in running the business.

In all, the Greenbury report contained some 20 recommendations, the key elements of which are summarised below:

- Remuneration committees should consist only of NEDs. This should avoid pay being determined by directors with a direct financial interest. Remuneration committees should:
  - publish an annual report giving full disclosure of all the elements (basic pay, bonuses, share options, pensions and so on);
  - relate incentives to demanding performance targets, in order to 'align directors' and shareholders' interests;
  - explain pay policy to shareholders and justify any unusual or exceptional awards;
  - have the committee chairperson attend an annual general meeting (AGM) to respond to shareholders questions.
- Long-term incentive schemes to be approved by shareholders.
- Discounted share options. No longer should directors be awarded share options at a discount to the prevailing market price.

The Hampel Report 1998

The Hampel Committee was created in 1995 to review implementation of the findings of the Cadbury and Greenbury Committees. The Hampel Committee published its report in 1998. Most of the recommendations in the earlier reports were then published in 1998 by the London Stock Exchange as *The Combined Code: Principles of Good Governance and Code of Best Practice*. The Combined Code (although redrafted since its original publication) is the current applicable code of best corporate governance practice for UK listed companies. The recommendations of Hampel were along similar lines and on similar issues to Cadbury. An important contribution made by the Hampel Report was the emphasis attributed to avoiding a prescriptive approach to corporate governance improvements and recommendations. The Cadbury Report highlighted the importance of focusing on the spirit of corporate governance reform, and Hampel reinforced this by stipulating that companies and shareholders needed to avoid a 'box-ticking' approach to corporate governance. The Hampel Report emphasised the need to maintain principles-based, voluntary approach to corporate governance rather than a more regulated and possibly superficial approach. This is typical of the UK approach to corporate governance and accounting as opposed to the US style of legislation, the rules-based approach

In some ways (such as the role of institutional investors in corporate governance), Hampel could be interpreted as being less demanding than Cadbury. Indeed, there is a widely held perception that the report represented the interests of the company directors more than those of shareholders and that much of the positive impact from the Cadbury Report was diluted by the Hampel Report. Certainly, in the area of corporate social responsibility and corporate accountability to a broad range of stakeholders, there was a significant change in fact between the Cadbury Report and the Hampel Report. The Hampel Report clearly felt the need to redress the balance between shareholders and stakeholders and made strong statements on these issues. For example, the Hampel Committee stated that:

> The importance of corporate governance lies in its contribution both to business prosperity and to accountability. In the UK the later has pre-occupied much public debate over the past few years. We would wish to see the balance corrected. Public companies are now among the most accountable organizations in the society... which strongly endorse this accountability and we recognize the contribution made by the Cadbury and Greenbury Committees. But the emphasis on the accountability has tended to obscure a board's first responsibility to enhance the prosperity of the business over time (Hampel Report, 1998: 7, para.1.1) (Solomon. and Solomon, 2004).

An important contribution made by the Hampel Report is related to pension fund trustees as they are the largest group of investors. Pension fund trustees were targeted by the report as a group who needed to take their corporate governance responsibilities more seriously. In particular, pension funds (and their trustees) were encouraged by the Hampel Committee to adopt a more long term approach to institutional investment, in order to avoid short-termism for which UK companies are notorious. Pension funds were highlighted as the main culprits in placing short-term pressure on their investing companies. This discussion in the Hampel Report has been instrumental in encouraging an overhaul in the pension fund trustee's role, culminating in the recent Minors Review of the trustee's role and responsibilities (Myners Report, 2001).

The impact of the Combined Code (and its predecessors) on UK company directors and institutional investors has been far-reaching, especially in the area of investor relations and shareholder activism. In a decade, corporate attitudes towards their core investors have been transformed from relative secrecy to greater transparency. Similarly, the attitudes of the institutional investors have been transformed from relative apathy towards their investee companies' activities to an active interest. As was the case for Cadbury and

Greenbury, the Hampel Report could also be seen as reactive rather than proactive, as further significant UK corporate failures arose from weak corporate governance structures between the publication of the Cadbury Report and the Hampel Report. One of these was a fall of the major UK bank, Barrings, which created shockwaves through the corporate and financial communities throughout the UK and, indeed, across the world.

The Turnbull Report 1999

The Combined Code (1998) dealt with internal control in Provisions D.2.1 and D.2.2. In these provisions, the code stated that company directors should conduct a review of the effectiveness of the internal control systems and should report this information to shareholders. The Turnbull Committee was established specifically to address the issue of internal control and to respond to these provisions in the Combined Code. The report provided an overview of the systems of internal control in existence in UK companies and made clear recommendations for improvements, without taking a prescriptive approach. The Turnbull Report was revolutionary in terms of corporate governance reform. It represented an attempt to formalise an explicit framework for internal control in companies. The aim was to provide companies with general guidance on how to develop and maintain their internal control systems and not to specify the details of such a system.

The Higgs Report 2003

Although the Cadbury Report and the Hampel Report stimulated substantial improvements in corporate governance in UK listed companies, certain areas have been highlighted for further examination. The fall of Enron spurred the UK and other countries into re-evaluating corporate governance issues, such as the role and effectiveness of NEDs. As evidenced from the Enron case, the NEDs were ineffective in performing their corporate governance role of monitoring the company's directors and were subject to conflicts of interest. Even though the emphasis on NEDs in the UK has represented an improvement in UK corporate governance, the UK government post-Enron felt obliged to set up an enquiry to examine their effectiveness.

The Higgs Report dealt specifically with the role and effectiveness of NEDs, making recommendations for changes to the Combined Code. The general recommendations included a greater proportion of NEDs on boards (at least half of the board) and more apt remuneration for NEDs. The report also concluded that stronger links needed to be established between NEDs and companies' principal shareholders. This would help to foster more effective monitoring of the notorious agency problem, as it would enhance the abilities of NEDs to represent shareholder interest and align in the interest of shareholders and directors. One important recommendation of the Higgs Report was that NEDs should assume chief responsibility as a champion of shareholder interest.

The Smith Report 2003

The UK Government in response to the Enron scandal commissioned this committee, *inter alia,* with the aim of examining the role of the audit committee in UK corporate governance. This report was published in 2003. The main issue dealt within the report concerned the relationship between the external auditor and the companies they audit, as well as the role and responsibilities of companies' audit committees. The creation of audit committees was a recommendation of the Cadbury Report and represented a clear means of monitoring company directors' activities. In the case of Enron, the failure of the audit committee as an internal audit function was one of the principal causes of the company's collapse. Improvements in this area represent one way of keeping a check on the production of reliable and honest accounts. Nevertheless, some have suggested that the report has not gone far enough. It has been suggested that a more prescriptive approach would have been preferable, which would, for example, prevent auditing companies from offering from other professional services, such as consultancy or IT (information technology) services or to client companies that they audit. However, the Smith Report preserved the UK tradition of a principles-based approach, attempting not to create a 'once size fits all' set of rules for listed companies. This would be counterproductive as not all companies would be in a position to comply.

Redraft of the Combined Code 2003

In July 2003, the Financial Reporting Council approved a new draft of the Combined Code, as intended from the Higgs Report in January 2003. Although the redrafted code was not as prescriptive as Higgs original recommendations, it retained much of the flavour of his concerns. Indeed, the redrafting was welcomed by both the corporate and institutional investment communities, despite their initial reactions to the Higgs Report. The revised code in fact retained almost all of the 50 recommendations contained in Higgs' original report. The language and the message were altered. The main reforms of the new code included the following:

- At least half the board of directors should comprise independent NEDs.
- A company's chief executive should not become chairman of the same company, except in exceptional circumstances.
- The board's chairman should be independent at appointment.
- A senior independent director should be appointed to be available to the company's shareholders, if they have unresolved concerns.
- Boards should undertake a formal and rigorous evaluation of their own performance, considering especially the performance and effectiveness of its committees and individual directors
- Institutional investors should avoid box-ticking when assessing investee companies' corporate governance.

- Companies should adopt rigorous, formal and transparent procedures when recruiting new directors.
- NEDs should only be reappointed after six years service, following 'a particular rigorous review.'
- NEDs can only continue after nine years service following annual re-elections and should be considered no longer independent.
- Boards should not agree to a full-time executive director accepting more than one non-executive directorship, or chairpersonship, in top 100 companies.

One of the main targets of the redrafted code was to readdress executive remuneration, as the new version of the code focused on forcing companies to avoid excessive remuneration, which displayed little relation to corporate performance. The revised Code also placed an emphasis on shareholder activism as a means of furthering corporate accountability and transparency.

The Sarbanes-Oxley Act 2002

In 2002, Paul Sarbanes, a Democrat Senator, and Michael Oxley, a Republican Congressman, were responsible for a radical piece of corporate legislation, the Sarbanes-Oxley Act. This act of congress is regarded by many commentators as the single most important piece of legislation affecting companies since the Securities Exchange Act of 1934. In the US, corporate crisis associated with companies such as Enron, Tyco and Global Crossing seems to have hastened the introduction of the Sarbanes-Oxley legislation. There is some evidence that the bankruptcy of WorldCom on 21 July 2002, and the public outrage that followed, encouraged President G.W. Bush to sign into law nine days later the Sarbanes-Oxley legislation (*New York Times,* 2 August 2002: 1).

The Sarbanes-Oxley Act introduced sweeping corporate law changes relating to financial reporting, internal accounting controls, and personal loans from companies to their directors, whistle blowing and destruction of documents. In addition, Sarbanes-Oxley severely restricts the range of additional services that an audit firm can provide to a client. There are increased penalties for directors and professionals who have conspired to commit fraud. Some examples follow of its provisions. Section 906 of the act requires that all periodic reports containing financial statements by the chief executive officer (CEO) and chief financial officer (CFO) of the company, certify that the report fully complies with the Securities Exchange Act and fairly present, in all material respects, the financial condition and results of operations. The penalties for knowingly certifying a statement, which does not comply with the requirements, can be severe: up to $1 million in fines and/or up to 10 years imprisonment. Section 1102 provides that 'knowing and willful' destruction of any record or document with intent to impair an official proceeding carries fines and/or imprisonment up to 20 years. Section 806 provides protection for employees who provide evidence of fraud. There is also protection for 'whistleblowers' in publicly traded corporations. No company, officer or employee may threaten or harass an employee who reasonably believes that a criminal offence has been committed. Section 501 of the legislation also aimed to promote rules to address conflicts of interest where analysts recommend securities when their companies are involved in investment banking activities.

The Sarbanes-Oxley legislation also established a Public Company Accounting Oversight Board (PCAOB) to be responsible to the Securities and Exchange Commission (SEC) for the regulation of auditing in US companies, inspection of accounting firms and disciplinary proceedings. As a result of the Sarbanes-Oxley legislation, some companies felt that the burden of compliance was too high in relation to the perceived benefits. Companies were reported to be spending "millions of dollars revamping their internal controls, updating compliance regimes, writing codes of ethics, setting up hotlines for internal complaints, writing governance principles and board committee charters. They are paying auditors and lawyers' greater fees, as well as directors" (*Financial Times*, 14 June 2004: 26). The chief executive of the New York Stock Exchange (NYSE), John Thain, argued that the additional burden of compliance was dissuading foreign companies from listing on the NYSE (*Financial Times*, 14 June 2004: 26).

Indian Committees

The Kumarmangalam Birla Committee on Corporate Governance (2001)

The Securities and Exchange Board of India (SEBI) had constituted a Committee on 7 May 1999 under the chairmanship of Shri Kumarmangalam Birla, the then-member of the SEBI Board "to promote and raise the standards of corporate governance." Based on the recommendations of this committee, a new clause (49) was incorporated in the Stock Exchange Listing Agreements (Listing Agreements). The first formal committee was appointed by SEBI, under the chairmanship of Kumaramanagalam Birla (known as Birla Committee). This was set up after the Confederation of Indian Industry (CII) code on corporate governance was framed to study the corporate governance from listed companies' perspective and culminated when its recommendations were included in the listing agreement. The recommendations were applicable to listed companies, their directors, management, employees and professionals associated with such companies and other bodies corporate.

The major recommendations of Birla Committee on corporate governance were:

- The board of directors of a company should have an optimum combination of executive and NEDs with not less than 50 per cent of the board consisting of NEDs. In case the company has a non-executive chairman, at least one-third of the board should consist of independent directors.

- Board meetings should be held at least four times in a year with a maximum time gap of four months between any two meetings.
- The board should set up a remuneration committee to determine the company's policy on specific remuneration packages for executive directors.
- The board should set up a qualified and independent audit committee.
- Companies should be required to give consolidated accounts in respect of all their subsidiaries. A company having multiple lines of business should be segmental reporting.
- A management discussion and analysis report should form part of the annual report to the shareholders covering industry structure, opportunities and threats, segment-wise or product- wise performance, outlook, and risks.
- Companies should arrange to obtain certificates from their auditors regarding compliance of corporate governance provisions and the certificates should be sent to stock exchanges and all the shareholders.

As mentioned, these recommendations were incorporated in the listing agreement (Clause 49) and were sought to be implemented within a time frame of three years. Later, these recommendations got statutory recognition when they were introduced provisions in the Companies (Amendment) Act, 2000.

Financial Reporting and Disclosures

Financial disclosure is a critical component of effective corporate governance. SEBI set up an Accounting Standards Committee, as a Standing Committee, under the chairmanship of Shri Y.H. Malegam with the following objectives:

- to review the continuous disclosure requirements under the listing agreement for listed companies;

- to provide input to the Institute of Chartered Accountants of India (ICAI) for introducing new accounting standards in India; and
- to review existing Indian accounting standards, where required, and to harmonise these accounting standards and financial disclosures on par with international practices.

SEBI has interacted with the ICAI on a continuous basis in the issuance of recent Indian accounting standards on areas including segment reporting, related party disclosures, consolidated financial statements, earnings per share, accounting for taxes on income, accounting for investments in associates in consolidated financial statements, discontinuing operations, interim financial reporting, intangible assets, financial reporting of interests in joint ventures and impairment of assets. With the introduction of these recent Indian accounting standards, financial reporting practices in India are almost on par with international accounting standards.

Implementation of Corporate Governance Requirements

The recommendations were implemented through Clause 49 of the Listing Agreements, in a phased manner by SEBI. They were made applicable to all companies in the BSE 200 and S&P CNX Nifty indices, and all newly listed companies, as of 31 March 2001. The applicability of the recommendations was extended to companies with a paid up capital of ₹ 100 million or with a net worth of ₹ 250 million at any time in the past five years, as of 31 March 2002. In respect of other listed companies with a paid up capital of over ₹ 30 million, the requirements were made applicable as of 31 March 2003. The accounting standards issued by the ICAI, which are applicable to all companies under sub-section 3A of Section 211 of the Companies Act, 1956, were specifically made applicable to all listed companies for the financial year ended on 31 March 2002, under the Listing Agreements.

Compliance with the Code and SEBI's Experience

In terms of SEBI's Circular No. SMD/Policy/CIR-03/2001 dated 22 January 2001:

- All companies are required to submit a quarterly compliance report to the stock exchanges within 15 days from the end of a financial reporting quarter. The report has to be submitted either by the compliance officer or by the CEO of the company after obtaining due approvals. SEBI has prescribed a format in which the information shall be obtained by the Stock Exchanges from the companies. The companies have to submit compliance status on eight sub-clauses namely:
  - Board of Directors:
  - Audit Committee;
  - Shareholders/Investors Grievance Committee;
  - Remuneration of Directors;
  - Board Procedures;
  - Management;
  - Shareholders; and
  - Report on Corporate Governance.

Stock exchanges are required to set up a separate monitoring cell with identified personnel, to monitor compliance with the provisions of the recommendations. Stock exchanges are also required to submit a quarterly compliance report from the companies as per the Schedule of Implementation. The stock exchanges are required to submit a consolidated compliance report within 30 days of the end of the quarter to SEBI. Both the Mumbai and National Stock Exchanges have submitted a consolidated quarterly compliance report for the quarter ended on 30 September 2002. It was observed that 1,848 and 741 companies were required to comply with the requirements of the Code, for the Mumbai and National Stock Exchanges, respectively. Of these, compliance reports were submitted in respect of 1,026 and 595 companies, for the Mumbai and National Stock Exchanges, respectively. The status of compliance with respect to provisions of corporate governance analysed from data submitted by the Mumbai Stock Exchange for the quarter ended on 30 September 2002 is set out below:

- Board of Directors;
- Audit Committee;
- Shareholders;
- Grievance Committee; and
- Remuneration Committee Board Procedures.

Management Shareholders Report on Corporate Governance

The key observations contained in the consolidated compliance report sent by the Mumbai and National Stock Exchanges are set out below:

- The compliance level in respect of requirements relating to board of directors, audit committee, shareholders, grievance committee and shareholders is very high.
- Many companies are yet to comply with the requirements relating to remuneration committee (which is not mandatory), board procedures, management and report on corporate governance.
- Few companies have submitted that the provisions relating to management and board procedures are not applicable.

SEBI observed that the compliance with the requirements in clause 49 of the Listing Agreement is, by and large, satisfactory; however, an analysis of the financial statements of companies and the report on corporate governance discloses that their quality is not uniform. This is observed on parameters such as the nature of qualifications in audit reports, the quality of the corporate governance report itself (which is often perfunctory in nature), and the business transacted and the duration of audit committee meetings. Variations in the quality of annual reports, including disclosures, raise the question whether compliance is in form or in substance, and emphasise the need to ensure that the laws, rules and regulations do not reduce corporate governance to a mere ritual. This question has come under close scrutiny in recent times. SEBI has analysed a few recently published annual reports of companies to assess the quality of corporate governance. The directors' reports could be classified into the following categories:

- Reports where there is no mention about the compliance with corporate governance requirements.
- Reports that state that the company is fully compliant with clause 49 of the Listing Agreement, but where independent auditors have made qualifications in their audit reports.
- Reports that mention areas of non-compliance with clause 49 of the Listing Agreement and provide explanation for non-compliance.
- Reports that mention areas of non-compliance with clause 49 of the Listing Agreement but provide no explanation for auditor's qualification or for reasons for non-compliance.

SEBI also observed that there is a considerable variance in the extent and quality of disclosures made by companies in their annual reports.

Rationale for a Review of the Code

SEBI believes that efforts to improve corporate governance standards in India must continue. This is because these standards are themselves evolving, in keeping with market dynamics. Recent events worldwide, primarily in the US, have renewed the emphasis on corporate governance. These events have highlighted the need for ethical governance and management, and for the need to look beyond mere systems and procedures. This will ensure compliance with corporate governance codes, in substance and not merely in form. Again, one of the goals of good corporate governance is investor protection. The individual investor is at the end of a chain of financial information, stretching from corporate accountants and management, through boards of directors and audit committees, to independent auditors and stock market analysts, to the investing public. Many of the links in this chain need to be strengthened or replaced to preserve its integrity. SEBI, therefore, believed that a need to review the existing code on corporate governance arose from two perspectives: (1) to evaluate the adequacy of the existing practices; and (2) to further improve the existing practice.

Naresh Chandra Committee (2002)

The initial stimulus for corporate governance reforms came after the South-East and East Asian crisis of 1997-98. Governments, multilateral institutions, banks and companies recalled that the devil lay in the details—the nitty-gritty of transactions among companies, banks, financial institutions and capital markets; corporate laws, bankruptcy procedures and practices; the structure of ownership and crony capitalism; stock market practices; poor boards of directors with scant fiduciary responsibility; poor disclosures and transparency; and inadequate accounting and auditing standards.

Following the Enron fiasco and subsequent enactment of Sarbanes-Oxley Act in the US, Government of India (GoI) (Department of Company Affairs [DCA]) had set up another committee to study corporate governance. This committee was formed under the chairmanship of Naresh Chandra (known as Naresh Chandra Committee/NC Committee).

This committee examined various governance issues, such as:

- Statutory auditor-company relationship including independence of audit functions and restriction on non-audit services.
- Need for rotation of statutory audit firms.
- Advantages of setting up an independent regulator.
- Role of independent directors for their composition in board.

India has not been in the middle of this global and Asian reform movement, as a reaction to corporate and financial crises. First, unlike South-East and East Asia, this movement did not start because of a national or region-wide macroeconomic and financial collapse. Indeed, the Asian crisis barely touched India. Secondly, unlike other Asian countries, the initial drive for better corporate governance and disclosure, perhaps as a result of the 1992 stock market 'scam,' and the onset of international competition consequent on the liberalisation of economy that began in 1990, came from all-India industry and business associations, and in the DCA. Thirdly, from April 2001, listed companies in India needed to follow very stringent guidelines on corporate governance, which rank among some of the best in the world. Sadly, there is a wide gap between prescription and practice. Worse, adverse legal consequences, for the defaulters, almost always get caught in the web of inefficiency, corruption and the intricate, dilatory legal system. Thus, while corporate governance reforms in India far outstrip that of many other countries, the performance in either lags very much behind.

After the Enron debacle of 2001, came other scandals involving large US companies such as WorldCom, Qwest, Global Crossing, and the auditing lacunae that eventually led to the collapse of Andersen. These scandals triggered another phase of reforms in corporate governance, accounting practices and disclosures—this time more comprehensive than ever before. In July 2002, less than a year from the date when Enron filed for bankruptcy, the Sarbanes-Oxley Bill was enacted. The act brought with it fundamental changes in virtually every area of corporate governance and particularly in auditor independence, conflicts of interest, corporate responsibility, enhanced financial disclosures, and severe penalties, both fines and imprisonment, for willful default by managers and auditors. It is fair to predict that the Sarbanes-Oxley Act will do more to change the contours of board structure, auditing, financial reporting and corporate disclosure than any other previous law in US history.

On 21 August 2002, the DCA under the Ministry of Finance and Company Affairs appointed this high level committee to examine various corporate governance issues. Among others, this committee has been entrusted to analyse and recommend changes, if necessary, in diverse areas such as:

- the statutory auditor-company relationship, so as to further strengthen the professional nature of this interface;
- the need, if any, for rotation of statutory audit firms or partners;
- the procedure for appointment of auditors and determination of audit fees;
- restrictions, if necessary, on non-audit fees;
- independence of auditing functions;
- measures required to ensure that the management and companies actually present 'true and fair' statement of the financial affairs of companies;
- the need to consider measures such as certification of accounts and financial statements by the management and directors;
- the necessity of having a transparent system of random scrutiny of audited accounts;
- adequacy of regulation of chartered accountants, company secretaries and other similar statutory oversight functionaries;
- advantages, if any, of setting up an independent regulator similar to the Public Company Accounting Oversight Board in the Sarbanes-Oxley Act, and if so, its constitution;
- the role of independent directors, and how their independence and effectiveness can be ensured.

Table 3.1

Summary of the Committees

S.No	*Committees*	*Year*	*Purpose*	*Focus*
1	Cadbury	1992	The Cadbury Committee was the first committee to be constituted to report on the financial aspects of corporate governance.	The Cadbury Report focused attention on the board of directors as being the most important corporate governance mechanisms, requiring constant monitoring and assessment. However, the accounting and auditing functions were also shown to play an essential role in good corporate governance, emphasising the importance of the corporate transparency and communication with shareholders and other stakeholders.
2	Greenbury	1995	Remuneration of directors	The Greenbury Committee was keen to ensure that directors' remuneration was linked to company performance, and the committee did not seem to see a problem with high levels of pay per se, as long as they were justified on the basis of the company's financial results.
3	Hampel	1998	To review implementation of the findings of the Cadbury and Greenbury Committees	The Hampel Report emphasised the need to maintain principles-based, voluntary approach to corporate governance rather than a more regulated and possibly superficial approach.
4	Turnbull	1999	The Turnbull Committee was established specifically to address the issue of internal control and to respond to these provisions in the Combined Code.	The aim was to provide companies with general guidance on how to develop and maintain their internal control systems and not to specify the details of such a system.

(contd...)

(contd...)

5	Higgs	2003	The Higgs Report dealt specifically with the role and effectiveness of NEDs, making recommendations for changes to the Combined Code.	The general recommendations included a greater proportion of NEDs on boards (at least half of the board) and more apt remuneration for NEDs.
6	Smith	2003	In response to the Enron scandal, the UK Government commissioned this committee, *inter alia*, with the aim of examining the role of the audit committee in UK corporate governance	The main issue is dealt with in the report concerned the relationship between the external auditor and the companies they audit, as well as the role and responsibilities of companies' audit committees
7	Sarbanes-Oxley Act	2002	In the US, corporate crisis associated with companies such as Enron, Tyco and Global Crossing seems to have hastened the introduction of the Sarbanes-Oxley legislation. There is some evidence that the bankruptcy of WorldCom on 21 July 2002, and the public outrage that followed, encouraged President G.W. Bush to sign into law nine days later the Sarbanes-Oxley legislation (*New York Times,* 2 August 2002: 1).	The Sarbanes-Oxley Act introduced sweeping corporate law changes relating to financial reporting, internal accounting controls, and personal loans from companies to their directors, whistle blowing and destruction of documents. In addition, Sarbanes-Oxley severely restricts the range of additional services that an audit firm can provide to a client.

Source: Duppati and Mishra, 2008.

References

Conyon, M.J. and C.A. Mallin (1997). "A Review of Compliance with Cadbury", *Journal of General Management* 2: 24-37.

Duppati, G, and R.K. Mishra (2008). *Corporate Governance: Theory and Practices.* New Delhi: Excel Publications.

Myners Report (2001). *Institutional Investment in the United Kingdom: A Review.* London: HM Treasury.

Solomon, J. and A. Solomon (2004). *Corporate Governance and Accountability.* Chichester: Wiley.

4

International Perspective of Oil Trends

Introduction

Oil and natural gas together with coal are the main fossil fuels. They not only are the major energy sources but are also the feedstocks for a great variety of man-made materials and products that range from gasoline and diesel oil to varied petrochemical and chemical products, including synthetic materials, plastics, and pharmaceuticals. What nature gave us as a gift, formed over the course of eons, is being used up rather rapidly. Fossil fuels continue to be significantly depleted and will become increasingly costly. Thus, there is a need to search for new sources and solutions.

All fossil fuels are mixtures of hydrocarbons, which contain varying ratios of carbon and hydrogen. Upon their combustion, they are irreversibly used up: carbon is converted into carbon dioxide (CO_2) and hydrogen into water. On a related note, the increase of the CO_2 content in the atmosphere is considered a major man-made cause of global warming. Coal is considered to have been formed during the so-called carboniferous period some 300 million years ago from the anaerobic decomposition of then-living plants. Its use for heating purposes began slowly and became more widespread in the 16th and 17th centuries mainly in England as a replacement for wood, which was becoming scarcer. Coal became dominant in the 18th century with the invention of the steam engine and the industrial revolution, which was also fuelled by coal, that followed. In the 19th and 20th centuries, coal continued to satisfy the ever-increasing energy demand of mankind, and today it is used primarily for generating electricity. Although the coal reserves may last for another two or three centuries, the mining of coal (except in areas suited for surface strip mining) is increasingly affected by socioeconomical, safety and environmental difficulties.

Oil and Natural Gas

Since the latter half of the 19th century, petroleum oil and natural gas have become increasingly important and dominant energy sources as well as raw materials for chemicals and man-made materials. They are certainly the most convenient fossil-fuel sources.

Much has been said over the years about the extent of our available oil and gas resources. At first glance, estimated world oil and gas reserves look impressive. These reserves seem not to have diminished in the last 50 years owing to continuous developments and improved technologies that have allowed exploration, production and recovery from increasingly difficult-to-access areas, such as the depths of the seas and inhospitable land areas from deserts to the Arctic. Conservation of energy and secondary and tertiary recovery from previously exploited fields are significant, but will not fundamentally change the long-range outlook.

Besides petroleum oil and natural gas (and also coal), we have additional sources of 'heavy' hydrocarbons, such as heavy-oil deposits in Venezuela, oil shales in various geological formations including the US Rocky Mountains, and vast tar sand deposits in the Canadian province of Alberta. The relatively stable hydrates of methane, as found under the Siberian tundras and along the continental shelves of the oceans, represent significant possibilities for natural gas deposits in the future. They will all eventually be exploited, although the difficulties and expense involved are extensive.

Besides the size of oil reserves (1-1.5 trillion barrels or some 200 billion metric tonne), one must also consider the expanding world population, which exceeds 6 billion now and will probably approach 9-10 billion by the mid-21st century, as well as increasing standards of living and demand in countries such as China and India if one wants to predict how long the estimated oil reserves may last. Present estimates of our proven oil reserves reveal that they would last for some 40 years at the current rate of consumption. Natural gas reserves are comparable but somewhat larger. New developments and improved recovery methods, however, could extend these estimates significantly.

Alternative Energy Sources

To satisfy mankind's ever-increasing energy needs, one has to consider all feasible alternative energy sources. Hydro- and geothermal energies are used where nature makes it feasible, but no major new sources have been found. Energy from the sun and wind, and waves and tides of the seas all have real potential and are increasingly exploited, but their use on such a large scale so as to substitute fossil fuels is not likely to significantly affect our energy outlook in the foreseeable future.

Recent Trends in Oil/Energy

Global energy demand increased to 5.6 per cent in 2010 mainly due to the rebound from the economic crisis, but in 2011 global energy demand slowed down and increased by only 1.6 per cent in 2011. However Organisation for Economic Co-operation and Development (OECD) and non-OECD countries saw opposing growth trends. OECD countries saw a decrease of 1.9 per cent, mainly due to the Fukushima accident and a decrease in coal consumption, while non-OECD countries saw a significant increase of 4.3 per cent during the period. Preliminary figures indicated a similar situation in 2012 with global energy demand continued to grow although at a slightly higher pace (around 3 per cent) but a 1.3 per cent decrease in demand in OECD countries due to the impact of the shift to shale gas in the US and the continued impact of the Fukushima disaster on Japanese energy demand.

Production

In terms of energy production, fossil fuels accounted for 81.7 per cent of global energy production in 2011. From 2010 to 2011, production of natural gas increased by 3.1 per cent and coal by 5.6 per cent; oil production increased by only 1.3 per cent. Among non-fossil sources, nuclear (5.1% of total energy production) saw a decline of 6.3 per cent, largely due to the Fukushima accident, while hydro (2.3% of total production) increased by 1.4 per cent between 2010 and 2011. With a 1.3 per cent increase, biofuels, mainly due to the large part of solid biofuels, (fuelwood, agro-residues), kept their 10 per cent share of global energy production. Other renewable sources (e.g. solar, wind), despite representing only 1 per cent of total production, registered the most pronounced growth rates. In 2011, wind generation increased by 27 per cent, solar thermal generation by 22 per cent, and solar photovoltaic by 91 per cent.

At a regional level, the OECD and Asia remained the two largest energy producing regions, each accounting for about 29 per cent of global energy production in 2011. However, the annual change in production levels was much higher in Asia than in OECD; this was especially true in 2011 with opposing trends showing an increase in Asia (+5.6%) compared to a decrease of 0.6 per cent for OECD. The trends are presented in Tables 4.1 and 4.2. However, it is the Middle East which registered the largest growth among all regions (+8.9%).

Table 4.1

Primary Supply of Oil

(*Mtoe*)

Countries	*1971*	*1973*	*1980*	*1990*	*2000*	*2005*	*2009*	*2010*	*2011*
India	21.83	24.28	33.20	61.44	112.71	126.45	159.60	161.74	165.69
Brazil	28.15	37.94	55.64	58.89	88.23	87.11	95.42	104.73	109.03
China	39.62	52.40	88.92	118.78	221.08	317.65	374.25	428.25	442.25
Russia	-	-	-	263.78	126.11	129.20	137.92	139.08	158.52
OECD- US	824.33	934.32	954.97	920.09	1058.95	1139.76	1001.73	1003.58	983.90

Source: International Energy Agency Report, 2012.

Table 4.2

Production of Crude Oil and NGL (Natural Gas Liquids)

(*Mtoe*)

Countries	*1971*	*1973*	*1980*	*1990*	*2000*	*2005*	*2009*	*2010*	*2011*
India	7.46	7.35	10.74	35.32	37.24	37.14	42.82	43.17	42.91
Brazil	8.66	8.60	9.47	33.39	65.34	86.94	109.59	112.83	110.45
China	40.12	54.48	107.85	138.31	163.08	181.43	203.16	203.03	203.26
Russia	-	-	-	526.25	323.26	468.71	506.54	514.86	522.29
OECD-US	649.26	660.41	698.45	680.61	665.58	667.58	671.04	693.45	751.79

Source: International Energy Agency Report, 2012.

In terms of production at the country level, many countries experienced a major increase in their own domestic production in 2011. This is the case for instance for China with coal and natural gas production increasing by 9.3 per cent and 8.3 per cent, respectively, nuclear by 16.9 per cent, solar and wind together by 39.2 per cent, however hydro production saw a decrease of 3.2 per cent, mainly due to less precipitation. The growing production of energy in China is a consequence of the growing demand in the country.

Chinese oil fields are ageing, their reserves-to-production ratios (R/P ratio) are low, and domestic oil production is nearing its peak. As a result, the country is almost entirely dependent on the international oil market to meet incremental oil demand. China became a net oil importer in 1993. For the past 17 years, China has experienced strong economic growth, and recently became the second-largest economy in the world. Even during the recent financial and economic crisis, China managed to achieve 8.7 per cent GDP (gross domestic product) growth in 2009, and 10.3 per cent in 2010.

It requires a great amount of energy-intensive raw materials and infrastructure to satisfy China's expanding consumer demand, as well as the rest of the world's demand for Chinese manufactured goods. This has stimulated output from heavy industry, which also received a boost from China's recent stimulus spending on energy-intensive infrastructure and buildings. The associated demand for fuel used to transport goods and to provide the growing fleets of private vehicles (China became the largest auto market in the world in 2009), as well as the rising demand for petrochemical feedstocks, has kept upward pressure on oil consumption. China's great hunger for energy, in particular its strong oil imports, contrasts with the recent fall in demand exhibited by major industrialised countries, which were hit harder by the recession. According to the International Energy Agency (IEA) research, almost half of global oil demand growth in the next five years will come from China. Looking farther ahead, the scenarios in the IEA's *World Energy Outlook* 2010 shows China importing 79 per cent of the oil it consumes by 2030, and accounting for a larger increment in oil demand than any other country.

For instance, when world oil demand fell 1.24 million barrels per day (mb/d), China's rose by 0.7 mb/d. Similarly, while world gas demand fell by 2 per cent, China's gas demand increased by 11 per cent. As domestic production is at or near its peak, virtually every incremental barrel or cubic meter of oil or gas consumed must be imported.

Table 4.3

Production of Oil Products

(*Mtoe*)

Countries	*1971*	*1973*	*1980*	*1990*	*2000*	*2005*	*2009*	*2010*	*2011*
India	20.35	21.24	26.11	52.87	103.96	130.61	195.71	207.48	213.94
Brazil	26.94	30.01	55.70	61.73	85.14	91.87	95.90	96.44	98.68
China	32.40	41.77	78.37	108.18	199.93	285.72	361.65	409.32	423.42
Russia	-	-	-	269.72	179.91	208.85	231.61	244.11	252.48
OECD- US	720.50	806.47	896.13	911.96	1014.96	1048.51	989.38	1016.27	1019.88

Source: International Energy Agency Report, 2012

The role of non-OECD countries in the current world energy picture is becoming more and more prominent when ranking countries by total primary energy supply (TPES), as shown in Table 4.4. In 2011, China continued to outpace the US in terms of TPES, with China accounting for more than 20 per cent of global TPES, while the US accounted for only 17 per cent. India and the Russian Federation ranked third and fourth, respectively. Japan, the second largest OECD consuming country, is in fifth position, despite the Fukushima accident. It should also be noted that Korea overtook both France and Canada to become the eighth largest energy consuming country in the world.

Besides being the main energy producing region, the OECD is also the largest importer. In terms of self-sufficiency, the OECD is last among all regions, with a production/TPES ratio of 73 per cent, far behind the closest region, Asia, at 89 per cent. The OECD statistics are largely driven by the importance of the US, alone accounting for 41 per cent of regional TPES. Japan and Germany follow, with 9 per cent and 6 per cent of regional TPES. After a short period of decrease in the mid-1970s and early 1980s, the OECD energy supply has progressed regularly, with an average annual growth rate of 1 per cent since 1971. However, in 2011, TPES decreased by 1.9 per cent, primarily due to the Fukushima accident and decreased coal consumption, an opposing trend to the 1.9 per cent increase in GDP.

Table 4.4

TPES: Top 10 Countries in 2011

Country	*TPES (Mtoe)*	*Share in World TPES (%)*
People's Rep of China	2728	21
United States	2191	17
India	749	5.7
Russian Federation	731	5.6
Japan	461	3.5
Germany	312	2.4
Brazil	270	2.1
Korea	260	2.0
France	253	1.9
Canada	252	1.9
Rest of the World	4907	37
World	13114	100

Source: Compiled from International Energy Agency Report, 2012.

Since the oil shocks of the 1970s, the OECD has diversified its energy supply. Consequently, the fuel shares in TPES have evolved, as shown in Figure 4.2. Although oil remains the main component of TPES, its share decreased from 51 per cent in 1971 to 36 per cent in 2011. The decrease was compensated by a pronounced increase in nuclear (from 1% to 10%) and by an increased penetration of natural gas (from 20% to 25%). The trends in electricity generation are even more pronounced than those of TPES, as presented in Figure 4.2. Electricity production has almost tripled since 1971, increasing on average by almost 3 per cent per year, more than twice as fast as TPES, at a rate comparable to that of GDP. Oil has been almost completely displaced by the dramatic increase of nuclear (on average, almost 8% per year), and by the increase of natural gas. Hydro has progressed very slowly, losing 10 per cent in the share over the time period. Most importantly, coal has kept the dominant role as electricity source, with a share in 2011 (33%) comparable to that of 1971.

The role of coal in total final consumption (TFC) within OECD countries is much less important than that in electricity generation. In final consumption, oil is the key player. Oil consumption (in the form of secondary products like motor gasoline) decreased after the oil shocks of the 1970s, but since the mid-1980s it started to grow again. In 2011, oil consumption grew by more than one per cent and still accounted for about half of the total final consumption within the OECD.

Figure 4.1

Details of the TPES by Fuel

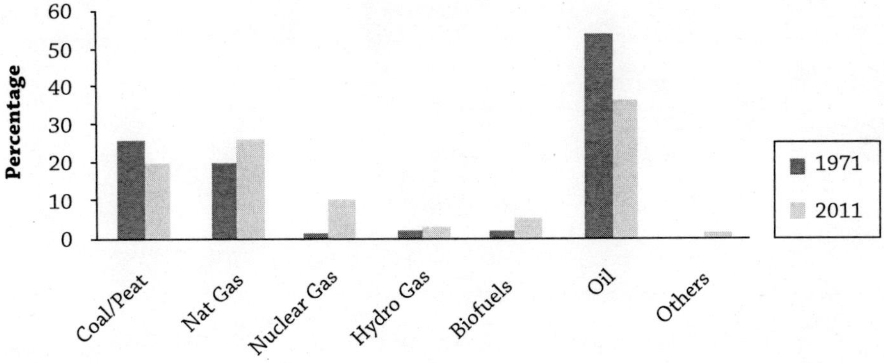

Source: Compiled from International Energy Agency Report, 2012

In 2011, energy production in non-OECD Americas was 3 per cent higher than in 2010, with a large variability between the fuels, for example, important changes were seen in Brazil (+13.4% for natural gas, +6.2% for hydro but -6% for biofuels) and Peru (+54% for natural gas, +85% for coal but -3.1% for oil). Variability of production was also common in Venezuela (-23% for coal, +1.1% for oil and +9% for hydro) and Argentina (+38.5% for coal, +16.5% for biofuels but -5.9% for oil).

Figure 4.2 shows that oil largely remained the dominant fuel in the supply (43% in 2011), followed by comparable shares of natural gas and biofuels (approximately 20%). While the share of natural gas in TPES increased from 9 per cent to 22 per cent between 1971 and 2011, the share of biofuels and waste decreased from 32 per cent to 19 per cent. Urbanisation and increases in the purchasing power of citizens explain the move away from traditional biofuels.

Figure 4.2

Non-OECD Americas: TPES by Fuel

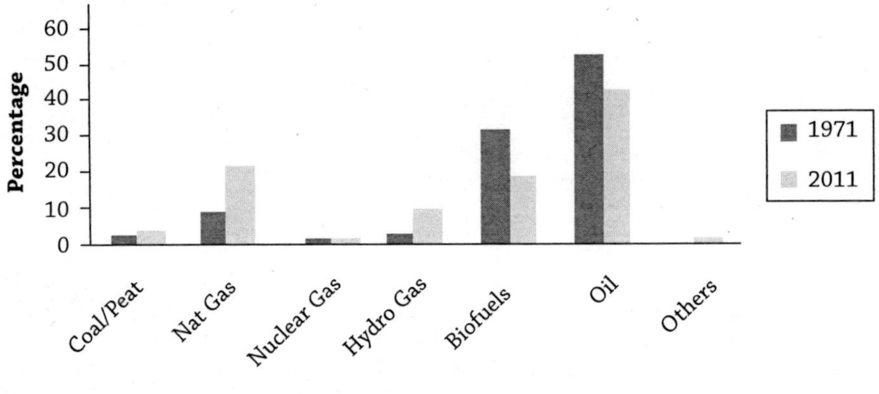

Source: Compiled from International Energy Agency Report, 2012

In 2011, Africa produced 8.4 per cent of the world's energy. African production is dominated by oil (39%), followed by traditional biomass (31%), natural gas (15%) and coal (13%). The use of biofuels (mainly fuel wood) is significantly higher across Africa than the world average. For commercial types of energy, production is unevenly distributed across sub-regions. Crude oil and natural gas tend to be concentrated in a few countries of North, West and Southern Africa. In 2011, Nigeria, Angola, Algeria and Egypt produced together 75 per cent of the crude oil of the region, while Algeria produced alone more than 40 per cent of the regional output of natural gas. Coal is produced almost exclusively in South Africa, the sixth coal exporter in the world, and a significant coal consumer itself.

Figure 4.3

Africas: TPES by Fuel

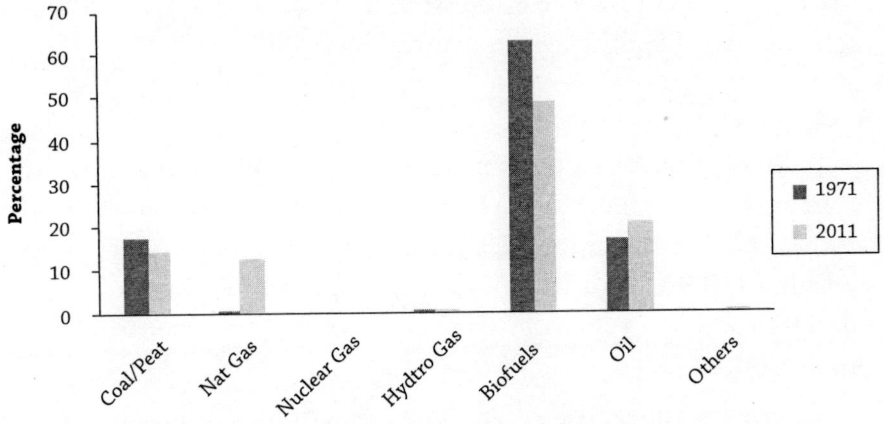

Source: Compiled from International Energy Agency Report, 2012

In 2011, Africa produced 13 per cent less oil than in 2010, mainly due to the Arab spring uprising in Libya, but still represented 11 per cent of world crude oil output. Africa exported 80 per cent of this production. Among the region's oil producers, output decreased for Libya (-70%), Nigeria (-1.8%) and Angola (-7.8%), while it increased for the new producer in the region, Ghana. Natural gas production decreased by 4.4 per cent with almost half of the output exported in 2011, both by pipeline and liquified natural gas (LNG). As shown in Figure 4.3, TPES in Africa is dominated by solid biofuels (mainly fuelwood), with a share in 2011 (49%) much higher than the world average (10%). The presence of large forests, agro-industry, agriculture, a large rural population and a low GDP per capita have resulted in a large use of solid biofuels for cooking. Because of the extensive use of wood and charcoal, energy intensity is higher than the world average.

However, the share of biofuels in TPES decreased significantly between 1971 and 2011, due to increased electrification, and particularly the recent development of power generation from natural gas. Natural gas increased its share in TPES from 1 per cent in 1971 to 13 per cent in 2011. Coal continued to represent an important share of African TPES (15% in 2011), due to the continued high reliance of South African demand on it (70% in 2011).

Asia

In 2011, the Asian total primary energy supply continued its pronounced growth (+6%), in line with the strong economic growth for the region. The region's GDP increased by 7.6 per cent, dominated by China (+9.3%) and India (+6.9%). Asia was the second largest energy-producing region in the world, with 29 per cent of global production.

Figure 4.4 shows the recent steep growth in production, led by China, the largest coal producer in the world. Despite this growth, the region as a whole is a net energy importer, as its internal demand is growing even faster than its production. Self-sufficiency has slowly declined over the last 20 years for both China (89% in 2011) and India (72% in 2011). Even Indonesia, which is still self-sufficient because of its large coal exports, became a net oil importer in 2004.

As shown in Figure 4.4, coal is the main energy source in the region, accounting in 2011 for more than half of the TPES, well above the world average of 29 per cent. Figure 4.5 shows that in 2011, the Asian coal supply increased by almost 9 per cent. Growth was driven by the 10 per cent increase of China, which consumed alone 49 per cent of the world's coal. In 2011, China also increased its oil supply by 3.3 per cent, becoming the world's second largest oil importer behind the US and just outpacing Japan.

Figure 4.4

Asia: TPES by Fuel

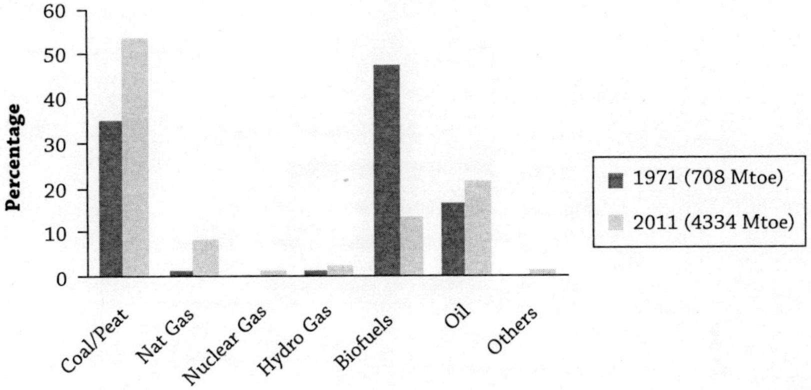

Source: Compiled from International Energy Agency Report, 2012

Increase in per capita GDP, urbanisation and electrification programmes have lowered the share of biofuels (predominantly solid biofuels) from 47 per cent in 1971 to 13 per cent in 2011 in the region. Due to the expansion of infrastructure, natural gas has become significant in the fuel mix (8%). In 2011, China itself increased natural gas production by 8.3 per cent, and also strongly increased its LNG imports that had started in 2006.

As shown in Figure 4.5, Asia has the largest share of coal in the electricity mix among all regions (69% in 2011). Among the largest electricity producing countries, coal provided 79 per cent of electricity in China, 68 per cent in India, 51 per cent in Chinese Taipei and 44 per cent in Indonesia. In 2011, total electricity generation in Asia increased by 10 per cent led by the 12 per cent increase of China. Even if electricity production grew in the region at an average annual rate of 8.4 per cent since 1971, its consumption per capita was still less than two thirds of the world average in 2011.

Figure 4.5

Share of Coal in Electricity Generation in 2011

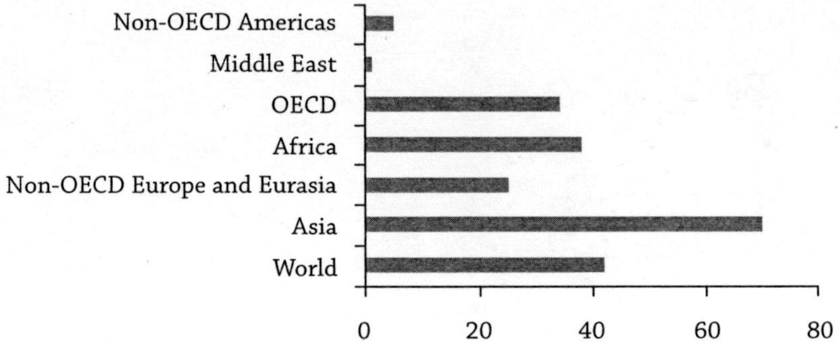

Source: Compiled from International Energy Agency Report, 2012.

Non-OECD Europe and Eurasia

Dominated by the Russian Federation (62% of regional TPES in 2011), Ukraine (10.8%) and Kazakhstan (6.6%), non-OECD Europe and Eurasia represent 9 per cent of the world energy supply. With a production/TPES ratio of 155 per cent, it is one of the most energy self-sufficient world regions.

Oil and natural gas are the largest contributors to energy production in the region (37% and 40% respectively in 2011), followed by coal and nuclear. Natural gas production increased in the largest producing country, the Russian Federation, with the 2011 output up 2.4 per cent from 2010. Crude oil production and nuclear power production were also up, 2 per cent each. In 2011, nuclear power provided 17 per cent of the electricity of the region, especially developed in Ukraine (46%), Armenia (34%) and Bulgaria (33%). However, it has been the development of natural gas in the region which is the most significant variation in the regional fuel mix over the last three decades. In 2011, oil, coal and natural gas still accounted together for more than 85 per cent of the supply.

In 2011, natural gas had the largest share in the regional TFC (31%), followed by oil (25%), heat (21%) and electricity (14%). Natural gas was also dominant in the regional electricity mix (40%) with coal as the second fuel of choice (24%). In 2011, coal- and oil-generated electricity increased (+6% and +74% respectively), while solar, geothermal and wind electricity generation more than doubled (+122%).

The energy profile of the non-OECD Europe and Eurasia region is largely influenced by major energy producers and exporters such as the Russian Federation. In 2011, the Russian Federation produced 10 per cent of global energy, 20 per cent of global natural gas, and 13 per cent of global oil. A key world energy exporter, the Russian Federation is increasing its production faster than its domestic supply. In 2011, net exports in the region were equivalent to 44 per cent of total production, compared to 32 per cent in 1990. Within the region, the Russian Federation produced 77 per cent of total oil, 76 per cent of total natural gas and 60 per cent of total coal in 2011, keeping its position as first exporter of natural gas, second exporter of crude oil and third exporter of coal in the world.

Since 2000 the economies of non-OECD Europe and Eurasia have been expanding at an average rate of 5 per cent per year. In 2011, they experienced a 4.2 per cent increase in GDP. In line with GDP growth, energy production has increased by 0.8 per cent per year on average since 1971 and remains one of the most energy intensive region among all regions, with a TPES/GDP ratio almost two times as large as the world average.

It is evident from the above that the OECD has diversified its energy supply from oil to biofuel and renewable energy. For instance, although oil remains the main component, its share decreased by 20 per cent from 1971 to 2011. The decrease was compensated by a pronounced increase in nuclear (from 1% to 10%) and by an increased penetration of natural gas (from 20% to 25%). The trends in electricity generation are even more pronounced than those of

TPES, as presented in Figure 4.3. Electricity production has almost tripled since 1971, increasing on average by almost 3 per cent per year, more than twice as fast as TPES, at a rate comparable to that of GDP. Oil has been almost completely displaced by the dramatic increase of nuclear (on average, almost 8% per year), and by the increase of natural gas. Hydro has progressed very slowly, losing 10 per cent in the share over the time period. Most importantly, coal has kept the dominant role as electricity source, compared to that of 1971. On the other hand the Non-OECD countries it is different. The diversification is insignificant.

According to IEA,[1] much observation regarding the overseas activity of China's national oil companies (NOCs) has presumed that the firms are acting under instructions and in close co-ordination with the Chinese government. Some experts have also expressed concerns that the activities of NOCs could result in reduced and more expensive supplies to other oil-importing nations. The IEA has carried out an assessment based on original research and secondary sources which found that the relationship between the Chinese government and the NOCs is complex with often divergent interests. While China's NOCs are majority-owned by the government (domestic and overseas private shareholders own minority stakes for publicly listed subsidiaries), they are not government-run. Their observed behaviour is the result of a complex interplay between individuals and groups associated with the firms, and whose interests are not always aligned, and where commercial incentive is the main driver.

Considering the growing significance of emerging nations in terms of oil and natural gas consumption and production, the following section presents the national oil companies and their activities from the point of sustainable practices and also corporate governance.

1. Overseas Investments by China's National Oil Companies, OECD/IEA, 2011

CASE POINT[2]

Energy Policy of Five Asian Countries

Countries	*Objective/ Strategy*
China	Demand driven. To tap the investments opportunities globally.
India	Pricing is an issue. Development of the hydrocarbon industry.
Japan	The government's energy strategy plan encourages Japanese companies to increase energy exploration and development projects around the world to secure a stable supply of oil and natural gas.
Malaysia	Malaysia's oil and gas policy historically has focused on maintaining the reserve base to ensure long-term supply security while providing affordable fuel to its population.
Singapore	It involves capitalising on the strengths of multiple related sectors that form the thriving energy ecosystem in Singapore.

Singapore

- The oil and gas industry is a key pillar of Singapore's economy, accounting for 5 per cent of the country's GDP. In 2009, this industry contributed $57 billion ($43.4 billion) to Singapore's total manufacturing output.
- Singapore is one of the world's top three export refining centres and the world's third largest oil and oil products trading hub, making it the undisputed oil hub in Asia. Central to the oil industry's phenomenal success is Singapore's unique 'plug and play' infrastructure and complete supply chain integration.

China

- China, unlike Russia or promising countries such as Nigeria, Angola, Venezuela or the Caspian Sea Republics, offers no exciting or valuable potential in oil exploration and production (E&P) activities and accordingly is strategically more considered as a second-tier oil investment-driven partner.
- On the one hand, China's consistent growth in GDP has been fuelling increasing demand for oil which an ill-developing domestic

2. Compiled from IEA and World Bank reports.

industry, facing at best since the late 1990s a plateau in production capacity, is unable to meet.

- This lack of growth in the oil sector is related in part to the forced decline (3%-5% yearly) of its major oil production area of Daqing (representing about 30 per cent of its total production output), and in part to the limited and marginally over-compensating output delivered by emerging production from vast yet remote sources in China's west and some promising offshore fields.
- It is also due to the insularity of its industry, which has been unable to protect its resources from international influence and which has kept it from forging significant foreign alliances, by which the industry could progress through the implementation of up-to-date technologies or production-sharing stimuli partnerships.

Malaysia

- The Petroleum Development Act of 1974 enacted by the Malaysian parliament is the basis for the development of the oil and gas industry in the country. Since then, the industry has developed to become one of the most important economic sectors.
- Malaysia has the 25th largest oil reserves and the 14th largest gas reserves in the world. The total reserve is of the order of 18.82 billion barrels oil equivalent (boe), with a crude production rate of 600 thousand barrels per day. The average natural gas production stands at approximately 5.7 billion standard cubic feet per day.
- Malaysia is the second largest oil and natural gas producer in Southeast Asia, the second largest exporter of LNG globally, and is strategically located amid important routes for seaborne energy trade.
- Malaysia's oil reserves are the fifth highest in Asia-Pacific and one of the top 30 highest reserves in the world.

Japan

- Oil remains the most significant energy source in Japan, accounting for some 45 per cent of the country's TPES in 2011. Japan's oil demand steadily decreased from 5.71 mb/d in 1997 to 4.47 mb/d in 2010.

- Of the 4.8 mb/d of oil imported by Japan in 2012, 3.5 mb/d consisted of crude oil, 209 kb/d of natural gas liquids (NGLs) and feedstocks, and some 1.2 mb/d of refined products. About 83 per cent of Japan's crude oil imports in 2012 came from the Middle East. The country has 27 operational refineries with a total crude distillation capacity of around 4.5 mb/d.
- Japan has limited domestic energy resources. It meets less than 15 per cent of its own total primary energy use from domestic sources. It is the third largest oil consumer and importer in the world behind the US and China. Furthermore, it ranks as the world's largest importer of LNG and second largest importer of coal behind China.
- In the wake of the Fukushima nuclear incident, Japan's energy fuel mix has shifted, as natural gas, oil and renewable energy provide larger shares and supplant some of the nuclear fuel. Oil remains the largest source of primary energy in Japan, although its share of total energy consumption has declined from about 80 per cent in the 1970s to 43 per cent in 2011.
- The decline in oil use occurred as a result of increased energy efficiency and the expanded use of nuclear power and natural gas. Coal continues to account for a significant share of total energy consumption, although natural gas is increasingly important as a fuel source and is currently the preferred fuel of choice to replace the nuclear shortfall. Natural gas rose from 18 per cent in 2010 to 22 per cent of total primary consumption in 2011.
- Before the 2011 earthquake, Japan was the third largest consumer of nuclear power in the world, after the US and France, and nuclear power accounted for about 13 per cent of the country's total energy in 2010. In 2011, the nuclear energy share fell to 7 per cent of the total energy consumption.
- Hydroelectric power and other renewable energy comprise a relatively small percentage of total energy consumption in the country, although renewable energy is becoming increasingly significant as a fuel source.

- Japan has very limited domestic oil reserves, amounting to 44 million barrels as of January 2013, according to the *Oil and Gas Journal* (OGJ). Japan's domestic oil reserves are concentrated primarily along the country's western coastline. Offshore areas surrounding Japan, such as the East China Sea, also contain oil and natural gas deposits. However, development of these zones is held up by competing territorial claims with China.
- The two countries reached an accord in 2008 to jointly explore four natural gas fields and equally invest in the development of two fields—Chunxiao/Shirakaba and Longjing/Asunaro. Since the agreement was signed, the countries have continued unilateral actions in attempts to develop the gas fields. Tensions escalated with territorial claims by Japan in 2012 and China's installation of a production platform in the contested area in 2013.
- Consequently, Japan relies almost solely on imports to meet its oil consumption needs. Japan maintains government-controlled oil stocks to ensure against a supply interruption. According to the IEA, total strategic crude oil stocks in Japan were 590 million barrels at the end of December 2012, where 55 per cent of those were government stocks and 45 per cent commercial stocks.
- The Japan Bank for International Cooperation supports upstream companies by offering loans at favourable rates, thereby allowing Japanese companies to bid effectively for projects in key producing countries. Such financial support helps Japanese companies to purchase stakes in oil and gas fields around the world, reinforcing national supply security while guaranteeing their own financial stability.
- The government's goal in the 2010 Strategic Energy Plan is to import 40 per cent of the country's total crude oil imports from Japanese-owned concessions by 2030, up from an estimated 23 per cent in 2010.

As a result of the 2011 earthquake and a pressing need to secure energy supplies, Japan is promoting even more investment in overseas oil and gas operations as well as technology exchanges with various countries.

5

Corporate Social Responsibility and Environment Sustainability:

A Case Approach

This chapter presents the corporate solid responsibility (CSR) approach and environment sustainability approaches of the selected companies from the rapidly urbanising nations.

Recently, a comparative study on sustainability practices between New Zealand and India was undertaken by Collins *et al.* (2013). Their findings showed that, overall, the New Zealand companies surveyed were more engaged in both environmental and social practices compared to surveyed companies in India. Surveyed companies in India also were much more likely to issue an environmental or sustainability report. Previous research has shown that New Zealand, compared to international standards, has a very low reporting rate (Collins *et al.*, 2010).

In both countries, surveyed respondents were more engaged in social rather than environmental sustainability. The results relate to the literature on trade-offs. Social and environmental sustainabilities were not treated equally, but rather social sustainability practices were much more likely to be adopted. Why social sustainability practices are more common for business is an area that needs further research. The findings on drivers align with the literature (Mishra *et al.*, 2012), showing that the business case for sustainability has not found much traction in India. The complex relationship of costs and sustainability was one of the key findings of the study. Costs were simultaneously a driver and a barrier and impacted the business case for sustainability.

It will be interesting to see the cases of oil companies from this perspective.

CSR and Environmental Sustainability Practices in BRICS Countries

The most rapidly urbanising countries are Brazil, Russia, India, China and South Africa (BRICS). Industrialisation and economic growth are the two main market drivers that have put these countries amongst the most rapidly urbanising countries in the world and home to eight of the world's 15 largest cities in 2012.[1] Globally, the BRICS account for 40 per cent of the world's population, a quarter of its land mass and a fifth of the global GDP (gross domiestic product). China dominates the BRICS in terms of economic might with its GDP of US $13.4 trillion, 22 times the size of GDP in South Africa. While South Africa has the smallest economy of the BRICS, it is still a significant power within the Middle East and Africa region with a well-capitalised banking system, abundant natural resources and an established manufacturing base.

Unemployment levels are relativity low and falling amongst most of the BRICS with per capita disposable income rising in all of the markets over the 2008-2013 periods and every country experiencing an explosion of the middle class. The approach of the five countries towards CSR and environmental sustainability practices is reviewed below.

Brazil

Brazil's economy has not been weakened by financial crisis. It shows steady growth, as does companies' commitment to

1. Euromonitor International Ltd (2013). "Brazil, Russia, China, India, and South Africa Create Formidable Force for Economic Growth", *New Research Press Release.* Tuesday September 10.

sustainability reporting. Brazil now has the world's third highest number of sustainability reporting companies. In 2010, over 160 Brazilian Global Reporting Initiative (GRI) reports were registered on GRI's Reports List.

There are nearly 40 Brazilian companies in GRI's Organisational Stakeholder Program, and Brazilian experts are well represented in GRI's governance bodies. The Brazilian sustainability reporting audience has taken a leadership role through their involvement in GRI's Readers Choice Awards and conferences. Brazil's indigenous population and rich biodiversity are among the distinctive factors that influence the sustainable development field. A GRI National Annex is in development, ensuring that Brazilian issues and needs are captured in GRI's guidelines.

The upcoming 'Rio+20' United Nations Conference on Sustainable Development is also creating an opportunity. The conference is the latest development in Brazil's recent history of fostering organisational transparency and accountability. Financial institutions, electric utilities and state-owned companies are among those increasingly involved in the sustainability agenda. The largest players in these sectors, such as Banco do Brasil, Caixa Economica Federal, Banco Bradesco, Itau Unibanco, Itaipu Binacional, Electrobras or Petrobras, are all reporters. In the 1990s, a group of concerned Brazilian entrepreneurs founded the Ethos Institute for CSR. In 2000, the Ethos Institute launched the Ethos indicators, which help companies familiarise themselves with responsible business practices. In the same year, the São Paulo Stock Exchange, BM&F Bovespa, launched the Novo Mercado Index, which ranks companies according to their corporate governance performance.

The hosting of the 1992 United Nations Earth Summit in Rio was an early signifier of Brazilian commitment to sustainable development, and led to the establishment of important organisations in the field of sustainability. The first half of the 1990s saw the establishment of Greenpeace, the Institute of Ecological

Research, or the Group of Institutes, Foundations and Enterprises (GIFE) in Brazil.

Russia

Russian Union of Industrialists and Entrepreneurs (RSPP) is an independent NEO (non-governmental organisation). The Union has a membership base of over 120 regional alliances and industry associations representing key industries of the economy, including the fuel and energy industry, the machine-building industry, the investment-banking sector as well as the military industrial complex, the building industry, the chemical industry, and light and food industries. The RSPP has in its ranks more than 328,000 members representing industrial, scientific, financial and commercial organisations and individual members in all Russian regions.

RSPP's main lines of activity includes:

- Initiating new bill drafts and making continuous efforts to improve the existing legislation, including areas such as taxation, debureaucratisation of the economy, small and medium business development, liberalisation of foreign exchange regulation and the customs policy, legal, pension, and banking reforms, science and industry, investment policies, the natural monopoly reform and the agroindustrial complex.
- Maintaining regular contacts with authorities at the federal and regional level to keep them informed about the efficiency of applicable laws and to protect the interests of industrialists and entrepreneurs. Holding round tables, forums, conferences and public discussions on key issues of business development in Russia.
- Participating in business clubs, associations, committees and commissions to present the views of the business community, to resolve jointly the problems of cooperation between companies and to maintain effective contacts among RSPP members, including enterprises, industry associations and regional alliances.

- Developing regular bilateral and multilateral relations with business community organisations in other countries; provide assistance to Russian businessmen in developing their export capabilities, overcome discriminatory restrictions on export of goods and services from Russia and attract foreign investments for domestic production development.
- Coordinating, through the medium of International Council for Cooperation and Investment, efforts of Russian companies and companies with foreign capital in improving the Russian investment climate, especially in the spheres of technical regulation and licensing, customs regulation, taxation, capital markets, and regional investments.
- Facilitating efforts of Russian companies in their investment and trade activities abroad.
- Keeping the general public informed of the place and role of the Russian business in carrying out the prosperity-oriented reforms through public activities, including expositions, presentations and awards (Best Russian Enterprises, High Social Efficiency Organisation), and through on-going interaction with the mass media.
- Maintaining regular contacts with authorities at the federal and regional level to keep them informed about the efficiency of applicable laws and to protect the interests of industrialists and entrepreneurs.[2]

The Russian approach towords CSR and environment sustainability can be understood from the G20 Labour Ministers meet with the social partners on 19 July.[3] The details are presented below.

For the first time ever, on 18 July 2013, the joint meeting of the G20 Labour Ministers' meeting and the social partners entitled "Social Dialogue on Labour and Employment Priorities

2. *http://eng.rspp.ru/about*
3. *http://eng.rspp.ru/news/view/3235*

for Job Creation" was held in Moscow. The representatives of the international business community and labour unions shared with the G20 Labour Ministers their vision of the priorities for employment and labour market development policies.

The discussion was focused on two major topics, namely, the role of trade unions and that of the entrepreneurs in promoting economic growth, and creating quality jobs (including in small and medium enterprises). Within the above larger topics, different viewpoints were presented on issues such as reform strategies for growth and jobs, creating enabling environment for enterprises and entrepreneurship, improving employment *versus* fiscal consolidation. The delegates considered possible further steps aimed at increasing employability, reducing inequality and tackling informal employment, addressing demographic challenges, as well as enabling efficient labour activation policies, enhancing social protection and ensuring the implementation of the G20 commitments.

Opening the meeting, the Minister of Labour and Social Protection of the Russian Federation Maxim Topilin highly approved of the participation of the social partners in the G20 process and their contribution to solving the issues of labour and employment. He said, "These consultations are very useful and allow us to see a different perspective on the issues that we address. The social partners' ideas and recommendations based on their expertise facilitate the G20 decision-making process with respect to such key issues as employment, labour relations development and social protection." Maxim Topilin also got the delegates acquainted with the welcoming address by the President of the Russian Federation Vladimir Putin. Deputy Prime Minister of the Russian Federation, Olga Golodets emphasised the significance of the event and noted its high level of representativeness. She stressed, "Nowadays the world is becoming aware of the fact that economic and financial development is impossible in the absence of due attention to human capital, and without qualitative analysis of the problems of its development.

The Labour Ministers' Meeting is a result of the work done by each country in this sphere, and I hope that the documents that the Ministers eventually agree on, will bring us to the next phase of the development of labour relations jointly with the financial markets development".

Olga Golodets noted that the most vulnerable groups are those that are confronting the toughest challenges in terms of employment. Much attention is given to the problems of youth employment, employment of women with children and people with disabilities. Another topic of vital importance at the moment for all labour markets is the new quality of labour force supply. Qualification requirements are changing at a very high pace, and national systems of apprenticeship should keep up with the demand in the developing market. Finally, wage setting regulation is crucial. The Deputy Prime Minister noted the problem of speculative overstatement of salaries in some sectors and its artificial understatement in other. This issue is also one of the priorities, since the market of investment is very sensitive to level of salaries.

Chair of the Business 20 (B20) and the President of the RSPP Alexander Shokhin said that the B20 had prepared comprehensive recommendations on the most pressing issues of the international economic development, to be shared with the G20 governments and business community. "Creating enabling business environment should be accompanied by necessary changes of labour regulation standards, including new efficient forms of employment, as well as expanded opportunities in defining the type of the labour contract," Mr Shokhin pointed out. The business community is convinced that the small business has a significant role to play in the economic development and job creation. Therefore, the governments should provide an enabling environment for start-ups by launching consultation services, and ensuring availability of financing as well as entrepreneurship skills training for youth.

New high-tech jobs will not be created if there are no employees to work there. The B20 considers it necessary to provide people with broader access to high quality education systems. Finally, the employers supported the implementation of the activation policy oriented towards vulnerable groups that was pursued by Russia during its G20 Presidency. Chief of the Labour 20 (L20), Chairman of the Federation of Independent Trade Unions (FNPR) of Russia, Mikhail Shmakov reminded the participants that this Labour Ministers' meeting with social partners was the fourth episode of working in this format, which has obviously found its niche in the framework of the major forum of worlds' leading economies.

The L20 priorities are described in the L20 Statement, having been finalised at the L20 Summit recently held in Moscow. These priorities include creating new jobs, restoring growth and trust, strengthening active labour market programmes and actions for youth, ensuring inclusive growth and reducing inequality, creating decent jobs in SMEs (small and medium enterprises), delivering on past commitments and preserving hope for the future. "Many of these priorities coincide with the priorities of the Business 20," Mikhail Shmakov noted. The head of the FNPR specifically touched upon a number of recommendations, which trade unions are willing to hand over to the G20 Finance Ministers and heads of governments *via* the Labour Ministers. The L20 calls on the G20 countries to support aggregate demand and employment in view of a serious slowdown of economic growth; to put an immediate halt to austerity measures and corresponding cuts in public spending in the areas that provide social support and facilitate productive economic activity; to invest in infrastructure, education and quality public services including in the care economy. Likewise, the L20 considers it relevant to ensure that revenues are not leaking through tax havens; to maintain and increase the level of spending on active labour market programmes; to cooperate with the social partners to upscale apprenticeships and training programs and put in place a 'Jobs Guarantee' for young people; and to reverse the rise in income inequality by strengthening collective bargaining and set robust minimum wages to avoid further incomes decrease.

Additionally, trade unions are calling for improving the provision of affordable finance to SMEs and facilitating their insertion into global value chains, while ensuring that they can provide decent work to their employees. And, finally, the International Labour Organisation (ILO) supports the implementation of the commitment to adopt the global social protection floor and the ratification of the ILO Conventions on social security; establishing the employment targets or indicators to be used in the G20 Mutual Assessment Process, and empowering the G20 Employment Task Force with a permanent mandate to carry out regular monitoring of the implementation of the past agreements.

India

Following market-based economic reforms in 1991, India has become one of the growing economies, and is considered a newly industrialised country; however, it continues to face the challenges. For instance, the year 2012-13 was marked by slowing growth, lingering inflation pressures and risks from persisting twin deficits. Policy efforts helped contain fiscal deficits and moderate inflation towards the later part of the year. However, significant challenges remain ahead for reviving growth while maintaining macro-financial stability. In this context, it becomes important to address structural constraints to growth, improve governance, address asset quality concerns and correct external imbalances (Reserve Bank of India [RBI], *Annual Report*).

In spite of the reforms, India continues to face the challenges of poverty, illiteracy, corruption and inadequate public health. India's business and investment communities are beginning to recognise the benefits of sustainability reporting and organisational transparency (GRI).

this context, the 'Urban Foundation,' which was established in 1976 (today: National Business Initiative [NBI]), set as its goal the improvement of the quality of life of the black communities. Since 1994, political change and the efforts to balance out the unequal distribution of wealth from the times of apartheid have driven CSR forward in South Africa. The involvement today has its roots in the 'Black Economic Empowerment (BEE) Act' which was set up by the government in 2003 and specifies the advancement of historically disadvantaged groups in the population. Especially the 'Corporate Social Investment' guidelines of the BEE have a supporting effect since with their assistance, CSR programmes are formalised and the results can be communicated.

CSR: Its Emergence

In the late 18th century, a Scottish philosopher and economist named Adam Smith wrote numerous articles on these subjects, his magnum opus being *The Wealth of Nations* in which he espoused the concepts of free trade and the free market on which the classic market economy was based. Smith's principles were borne out. By the early 19th century, new technology saw jobs being created and living standards improved. Unchecked by regulation, businesses flourished and industrialists in Europe and the US amassed huge fortunes. However, few of these wealthy new industrialists were concerned about the well-being of their employees, society or the environment. The appalling conditions under which people worked were documented in the novels of Charles Dickens and inspired radical theorists such as Karl Marx and Friedrich Engels to write about new concepts on labour, socialism and communism. By the start of the 20th century, powerful corporations suffered a backlash against their widespread exploitation. Labour unions were formed, giving a voice to the workers, and governments began to assume more responsibility for welfare and infrastructure, gradually introducing anti-trust legislation. In the 1950s, emerging 'consumer power' saw companies beginning to develop interest in the social and human aspects of their markets—it was at about this time scientists and environmentalists started noticing some worrying changes to the environment.

The 1960s saw a shift in attitudes towards government and business. In 1962, Prof Milton Friedman, a Nobel Prize-winning economist, published his controversial *Capitalism and Freedom*. In it, he makes the case for economic freedom as a precondition for political freedom. The 1980s and 1990s saw communism collapse, globalisation emerge and the information technology revolution change the way the world did business. As globalisation intensified, so did environmental awareness and the emergence of responsible business practice. Key developments include: the Brundtland Commission, the formation of the World Business Council for Sustainable Development, and the United Nations Global Compact.

The cases of popular oil companies chosen from the fast growing emerging nations are presented below:

1. Case of LUKOIL.
2. Case of Petrobras.
3. Case of Ecopetrol S.A.
4. Case of Oil India Ltd. and
5. Case of Oil and National Gas Corporation (ONGC).

Case of LUKOIL

LUKOIL OAO, together with its subsidiaries, engages in the exploration and production of oil and gas, production of petroleum products and petrochemicals, and marketing of these outputs. The group has the following operating segments: exploration and production, refining, marketing and distribution, chemicals, power generation and other segment. The exploration and production segment explores for, develops and produces primary crude oil. The refining, marketing and distribution segment processes crude oil into refined products and purchases, sells and transports crude oil and refined petroleum products. The Chemicals segment refines and sells chemical products. The Power Generation segment produces steam and electricity, distributes them, and provides related services. The activities of the other business operating segment include business beyond the company's traditional operations. LUKOIL OAO was founded on 25 November 1991 and is headquartered in Moscow, Russia.

LUKOIL is a major vertically integrated international oil & gas company, whose crude oil accounts for 2.2 per cent of global output. Its strong position has been attained as a consequence of expanding the reserve base, increasing business scale by seeking out and executing strategic transactions for the past 20 years.

Company's Mission

The purpose of LUKOIL is to harness natural energy resources for human benefit. It aims to support long-term economic growth, social stability, prosperity and progress in the regions in which it operates. The company also has concerns for the environment and ensures sustainable use of natural resources. It aims to achieve consistent and long-term growth of the business, transforming LUKOIL into a leading global energy company. It also aspires to be a reliable supplier of hydrocarbons in the international energy market.

LUKOIL sets itself the objectives to create new value, maintain business stability, and provide shareholders with high return on their investments through asset value appreciation and cash dividends. LUKOIL aims to use all available means to achieve these objectives, including further efforts to reduce costs, increase operating efficiency, improve product and service quality, and utilise the latest technologies. LUKOIL is one of the world's biggest vertically integrated companies for production of crude oil & gas, and their refining into petroleum products and petrochemicals. The company is a leader in its core business in Russian and international markets.

LUKOIL Today

- The largest privately owned oil & gas company in the world by proved oil reserves.
- 0.8 per cent of global oil reserves and 2.2 per cent of global oil production.
- Fourth largest privately owned oil & gas company in the world by oil production.
- The Sixth largest non-state publicly traded oil company worldwide by production of hydrocarbons.
- The Thrid largest non-state publicly traded oil company worldwide by proven reserves of hydrocarbons.
- 16.6 per cent of Russian crude oil production and 17.7 per cent of Russian crude oil refining.
- One of the biggest Russian oil business groups with US $133 billion revenue and net income in excess of US $10 billion.
- One of the largest tax payers of the Russian Federation. The total amount of taxes paid in 2011 is US$38.4 billion.
- The company with second largest trading volumes among foreign companies traded on the London Stock Exchange (International order Book [IOB]) in 2011.
- A leader among Russian oil companies for openness and transparency. The first Russian company to receive full listing on the London Stock Exchange.
- The only private Russian oil company whose share capital is dominated by minority stakeholders.

LUKOIL producesd its first batch of automotive gasoline as per the Euro-5 standards. This was made possible by commissioning of a hydrofluoric alkylation unit as part of catalytic cracking complex at the Nizhny Novgorod Refinery. Euro-5 gasoline ensures so-called 'clean exhaust', reducing emission of sulphur oxides and the products of incomplete combustion of aromatics (including the particularly dangerous carcinogen and mutagenic, benzopyrene).

Corporate Governance Practices in LUKOIL

The board is responsible for general management of the Company's business except for the reserved matters of the General Shareholders Meeting (GSM). The key functions of the board include designation of the company's priority areas of the company's business, strategic, medium term and annual planning, assessment of performance results, preparation of the GSM, formation of the Management Committee as a collective executive body, approval of transactions in accordance with the existing legislation and the company charter, and other matters.

The 11-member board is elected by the GSM by cumulative voting (the candidates for whom the largest number of votes was cast are deemed elected). The company must include the election of the board members in the agenda of the annual GSM. Nominations to the board may be made by holders of at least two per cent of the voting shares within 30 days from the end of the fiscal year.

The board members are elected to serve until the next annual GSM and may be re-elected any number of times. The GSM may early terminate the office of all board members. Fees and compensations to all board members are established by the GSM. In its annual reports, the company includes information on the aggregate fees and compensations received by the board members during the year in question.

As regards the composition of the board, the company seeks to maintain the necessary balance between the experience, professional skills, knowledge of the company's business, as well as independence and objectivity in their judgements and decisions. As of the approval date hereof the board is comprised of the chairman of the board (who is a non-executive director), two executive directors (president of the company and the first executive vice president), and six independent directors (based on the criteria of the Corporate Governance Code of the Russian Federal Commission for Securities Market [FCSM]) and two non-executive directors.

The board activities are based on the board's work plan. There is a number of matters reviewed by the board on an annual basis in line with the Russian laws and the company's customary practices. For instance:

- In January, the board pulls together the preliminary operational results of LUKOIL Group for the past year and sets the goals for the current year and short-term goals, and in the course of this exercise, it identifies the priority areas of business for the current year and performs a comparative review of the company's operational performance for the recent years.
- In April, the board attends to the preparation of the GSM, and, in particular, it designates the record date, approves the agenda, the text and the forms of voting ballots; performs a preliminary review of draft resolutions on the agenda items; and handles other organisational issues.
- At its meeting in May, the board issues a preliminary approval of the Annual Report of OAO 'LUKOIL' for the past year which is then submitted to the GSM for approval, and summarises the results of its operations, hears the reports on the committees' annual performance and on execution of the board's instructions, and assesses its own performance during the year.
- Since 2011, the board approves the OAO 'LUKOIL' *Corporate Governance Report* in May.
- In June, at its first meeting, the newly elected board elects the chairman from among its members, and based on the recommendation of the board, the chairman appoints the secretary of the board and elects committee members.
- In July, the board forms the management committee of OAO 'LUKOIL' and determines the key terms and conditions of contracts to be entered into with its members. In furtherance of the Federal Law on joint stock companies, the board establishes the maximum remuneration of the auditor of

LUKOIL, and at the same meeting approves the work plan of the board for the period until the next annual GSM.

- In August, it summarises the preliminary results of LUKOIL Group's business for the first six months of the year and reviews budget and investment programmes performance in the current year.
- In November, the board sets the key parameters for the medium term plan of LUKOIL Group for the period of three years.

  Once every two years, the board approves the adjusted LUKOIL Group 10-year strategic development programme. In 2011, the board also approved a number of important documents to improve corporate governance, which are briefed below:

  – Regulations on Internal Control and Internal Audit at LUKOIL OAO and the Risk Management Policy of LUKOIL OAO.
  – Regulations on the Risk Management Committee.
  – Regulations on the organisation and conduct of control and audit reviews at LUKOIL OAO.

Apart from the above resolutions, the Board also reviews in more detail and developes an action plan for development of certain LUKOIL Group business units. This work is performed with a focus on the following issues:

- Benchmarking of the company's activities: R&D (research & development).
- On progress in the development of deposits in the Northern Caspian.
- On reforming the accounting system at LUKOIL Group companies.
- Progress report on the company's gas projects.
- On the use of new technologies and innovation for oil and gas exploration and production.

- To better familiarise the board members with the regions of the company's operation, the company has introduced the practice of offsite board meetings.
- In 2011, the meeting dedicated to the development of the North Caspian fields took place in Astrakhan. To be more specific, it was dedicated to the Progress Report on the North Caspian Fields Development; a decision was made to consider further implementation of the Caspian projects on top priority in terms of oil and gas short- and mid-term production enhancement.

In Compliance with the Corporate Governance Code in Russia

In accordance with point 7.2 of the Disclosure and Transparency Rules of the UK Financial Services Authority (FSA), as a foreign company with a standard listing in the UK, LUKOIL OAO has since 2011 been obligated to prepare a Corporate Governance Report for the previous year.

As a Russian corporation LUKOIL OAO (hereinafter, LUKOIL or the Company) conducts its business in compliance with the Corporate Governance Code (the Code) recommended in 2002 by the Russian FCSM for application by the Russian joint stock companies. The code is a voluntary standard that includes multiple provisions covering various areas of business of a joint stock company. FSFM identified the most important provisions of the Code which must be observed by the issuers seeking a listing at Russian stock exchanges. LUKOIL fully complies with these requirements, and its shares and Rouble bonds are the blue chips (A1 level) of the Russian Stock Exchange, MICEX. LUKOIL on a quarterly basis submits the reports on compliance with corporate governance standards to this stock exchange, and discloses its internal documents evidencing such compliance.

As regards the Code in general, LUKOIL includes in its annual reports—approved by the GSM—the detailed information on compliance with the Code using the form provided by the FCSM.

At present, LUKOIL is in full compliance with most of the Code's provisions; however, there is a number of exceptions. The main ones follow.

The company's board of directors have three committees under board committees. The Code recommends that a corporate conflicts resolution committee be created in a joint stock company in addition to such committees. The company does not presently have such a committee. However, the company does not exclude that such a committee may be set up in the future when there is a sufficient number of issues which the Code recommends to delegate to such committee. However, the company has long had units in charge of shareholder and investor relations.

The company's unit in charge of internal control and internal audit, which was functioning in 2011, reported to the company's President. All material documents of the company relating to internal control and internal audit are approved by the Board subject to prior recommendations from the Audit Committee. The competence of the Audit Committee also includes the assessment of the efficiency of the company's internal control procedures and development of improvement proposals, and performance efficiency assessment of the company's unit in charge of internal control and internal audit.

The secretary of the board, who in 2011 was head of the board office and member of the Company's Management Committee, acts as the corporate secretary of the Company. In the meantime, the majority of the organisational issues associated with preparation and procedures of the GSM, review of shareholders' applications, contacts with the registrar, are handled by the unit which is in charge of shareholder relations. This unit has efficiently performed these functions since the first years of the company's existence.

Moreover, in certain respects, the company striving to apply the best international practices exceeds the requirements of the Code. In particular, the Code recommends that the number of independent directors be at least one-fourth of the entire board. The Company annually assesses the independence of its directors in line with the Code, and in recent years more than half of the LUKOIL board have been independent. The Company has the *Code of Business Conduct and Ethics of OAO "LUKOIL"* which is a set of norms and rules of individual and collective behaviour of the Company's employees. This document addresses, in particular, issues such as relations between Company's employees, relations with the business partners, shareholders, governmental agencies and public organisations, environmental issues, industrial safety and prevention of conflicts of interests.

Environmental Policy

Company Policy

The LUKOIL Group strategic objective is to ensure dynamic sustainable development corresponding to that of global oil majors in terms of their efficiency and competitiveness. LUKOIL Group social responsibilities consist of an adequate working environment, appropriate remuneration, environmental safety and preservation of the cultural heritage. Being aware of its social responsibilities related to environmental protection and efficient use of natural resources, the company hopes the public will understand the complexity and the scale of the tasks it is facing.

The company's priorities are to expand production while securing occupational safety and protecting community and its employees health, and to preserve natural environment of the Kaliningrad Region and the south-east part of the Baltic Sea where the Company operates. To achieve its priorities, the company has set out the following goals:

- Improve health, safety and environmental management system.
- Use natural resources in efficient way.
- Minimise level of pollutants contained in waste waters and discharged into the air, and further perfection of industrial and consumption waste management.

- Prevent accidents and lower industrial injuries and accident rate.
- Provide safety and protect health of the company employees in the process of work.

To achieve these goals, the company has committed to:

- Ensure compliance with the federal, regional, and local laws, international liabilities assumed by the Russian Federation, as well as industry and corporate standards regulating company's activities in the area of industrial/occupational/ environmental safety.
- Take actions to reveal and mitigate workplace health risks; utilise natural resources so that to avoid environmental pollution. The present policy applies to all business units of the Company.

The Company's top priority objectives include creation of a safe working environment for its employees, health protection of LUKOIL Group personnel and local communities in the areas of LUKOIL Group operations, and environmental protection. To ensure accomplishment of the aforementioned tasks, the Company has developed and is successfully running its Health, Safety and Environment (HSE) Management System which includes compliance with fire safety and emergency response requirements. It is in line with the applicable Russian law, is based on the best Russian and global practices, and has been certified to comply with ISO (International Organization for Standardization) 14001 and OHSAS (Occupational Health and Safety Management System) 18001 standards.

The objectives of the HSE Policy (here in a after referred to as the policy) implemented by the open joint stock company 'oil company LUKOIL in the 21st century are to:

- Increase petroleum gas utilisation rate through expansion of gas-processing and power-producing capacities.

- Apply the zero-discharge principle while developing offshore fields.
- Increase the output of environmentally friendly fuels compliant with the European standards.
- Comply with greenhouse gas (GHG) reduction provisions of the Kyoto Protocol.
- Ensure efficient management of natural resources in the areas of LUKOIL Group operations and of those used for production by introducing power-efficient techniques and using alternative energy sources.
- Bring under control and gradually reduce both the amount and toxicity of emissions, discharges of pollutants and waste by introducing new cutting-edge techniques, equipment and materials and increased process control automation.
- Ensure continuous improvement of HSE performance, without limitation, through increased reliability of process equipment, its safe and accident-free operation, introduction of new technologies and automated emergency systems.
- Ensure that the management, personnel, rescue services and units of LUKOIL Group organisations are prepared to take accident elimination, fire fighting and emergency response measures, as well as enhance preparedness and provide more advanced equipment to fire-fighting and rescue units.
- Ensure more efficient development and implementation of the company programmes aimed at identifying and achieving the most critical HSE, occupational safety and emergency response objectives.
- Reduce anthropogenic environmental load resulting from the operation of newly commissioned facilities by ensuring better quality of front end and design documents.
- Exercise more efficient production control, corporate supervision and internal auditing to ensure compliance with the HSE regulations at LUKOIL Group sites based on cutting-

edge information technologies, technical diagnostics and remote monitoring techniques in line with ISO 14001, ISO 17020 and OHSAS 18001 international standards.

To achieve the above-specified goals, LUKOIL commits to the following:

- Implement the entire range of injury and accident prevention and mitigation measures that are practicable and available to it.
- Ensure that scheduled and implemented steps and measures related to prevention of impact upon personnel, population and environment are viewed as those of higher priority than emergency response measures targeting such impact.
- Ensure that there is continuous and consistent improvement in terms of how and to what extent the personnel, residents and environment are impacted by the company's operations, products and services, as well as consumption of natural resources in line with the current level of scientific, technological and social development.
- Continuously implement task-specific measures aimed at reducing oil and gas losses, as well as oil and gas processing losses, and preventing them from entering the environment.
- Introduce up-to-date scientific developments and technologies to gradually reduce specific consumption of natural resources, materials and energy and yet maintain the maximum possible production levels.
- Ensure compliance with the applicable Russian law, international treaties, industry-specific and corporate regulations, as well as national laws of the countries of LUKOIL Group operations governing health, fire and environmental safety and emergency response preparedness.
- Assess the impact that the scheduled business activities, products and services may have on the health of the personnel, population and environment.

- Assess production, occupational and fire-safety risks, develop and implement measures to mitigate and offset such unforeseen losses.
- Make and implement any managerial and production-related decisions that necessarily take account of substantial environmental issues, as well as production and occupational risks.
- Request that all personnel employed by LUKOIL Group perform their work in compliance with the applicable industrial, fire and environmental safety, health and emergency response rules and regulations, introduce respective incentives, provide training and advanced training to the company's personnel and LUKOIL Group employees.
- Make best efforts to improve environmental quality in the areas of LUKOIL Group operations.
- Request that corporate entities contracted by LUKOIL Group comply with the industrial, fire and environmental safety, health and emergency response regulations adopted by LUKOIL Group organisations.
- Publicly report to the Company's shareholders and personnel on industrial, fire and environmental safety, health and emergency response issues.
- Inform on a regular basis all parties concerned (public at large, executive and legislative bodies, etc.) regarding LUKOIL Group operations in the field of industrial, fire and environmental safety, and health and emergency response.
- Continuously improve the HSE Management System in compliance with ISO 14001 and OHSAS 18001 standards.
- Exercise control over compliance with the obligations assumed by the company as part of the Policy.

Caring for the Environment

A task-level environmental protection programme was developed in 2011 as part of the 2012-2021 LUKOIL Group Development

Strategy. The following strategic guidelines for environmental safety promotion were identified:

- Attain a 95 per cent level of utilisation of associated petroleum gas (APG).
- Discontinue discharges of contaminated waste waters into water bodies.
- Reduce emissions of GHGs and raise income through implementation of the mechanisms specified in Article 6 of the Kyoto Protocol.
- Complete elimination of 'previous environmental damages'.
- Do not exceed the ratio of one for generated waste to utilised waste.
- Do not exceed the 15 per cent limit for emissions above the allowable level in the payment structure for negative environmental impact.
- Lower the number of pipeline failures and reclaim lands contaminated due to these failures.

In its business, the company is guided by the highest HSE standards. Taking social responsibility for the efficient use of natural resources in all its earnestness and maintaining safe environmental conditions is regarded as one of the key factors to be considered over the course of project implementation. In 2011, the company and other organisations of the group took regular measures to ensure industrial and environmental safety, emergency prevention and emergency response. In July 2011, the company's Management Committee approved the 2011-2013 edition of the LUKOIL Group Environmental Safety Program for 2009-2013 developed on the basis of the updated version of the environmental aspects register, LUKOIL Group's proposals, as well as in view of the needs of the Power Engineering Business Sector, which has been separated in the corporate structure. The financing required for the programme implementation in 2011-2013 comes to around $2 billion. Over US $700 million was allocated for environmental safety measures in the reporting year. A considerable share of these funds was allocated for measures aimed at ensuring compliance with the state requirements to reach a 95 per cent level of utilisation of associated gas starting in 2012.

In 2011, the implementation of environmental measures helped reduce the majority of the environmental impact factors (including specific ones), with the exception of the atmospheric pollutant emissions. The growth of the emissions weight is accounted for by the considerable increase in the minor generation of electric power by the company's facilities and respective increase in the fuel consumption rate, as well as by the increased combustion volume of associated petroleum gas in a number of LUKOIL Group organisations. LUKOIL pays special attention to the utilisation level increase in order to ensure compliance with the Russian Federation (RF) Government requirement, i.e. attainment of at least a 95 per cent utilisation level of associated petroleum gas and minimisation of the above-norm environmental payments starting in 2012. All organisations of the Group are expected to be fully compliant with these requirements by 2015. According to the results of 2011, a twofold reduction in the discharge of contaminated waste water was registered, all due to the measures taken to enhance the efficiency of the treatment facilities. A 6 per cent reduction in water consumption is mainly due to the implementation of administrative and technical measures to ensure efficient water consumption in the organisations of the Power Engineering Business Sector (the group's largest water consumer). The environmental and economic mechanisms of the Kyoto Protocol served as an important driving force to invest in measures aimed at reducing the atmospheric emissions of GHGs. In general, as of the beginning of 2012, the portfolio of the Russian projects aimed to reduce emissions of GHGs in LUKOIL Group comprises 14 projects with a total volume of 33 million tonne of GHG emissions, in CO_2 equivalent, to be reduced in 2008-2012. Eight hydrocarbon projects of LUKOIL Group were approved by the RF Ministry of Economic

Development and Trade as joint-implementation projects under Article 6 of the Kyoto Protocol. The first revenue tranches from the sales of ERU (emission reduction units) with a total volume of € 1.3 million have already been received for three of the projects (with OAO RITEK as the operator). The revenue from ERU sales under other projects approved at the national level is expected in July/August 2012. For six of the projects, packages of documents, required for applications to be submitted to the Sberbank of Russia for approval, have been prepared. LUKOIL Group organisations ensured environmental safety in 2011 through a number of measures aimed at the efficient use of water resources, prevention of contamination of water bodies, reduction of pollutants in the atmospheric emissions, utilisation of accumulated oil waste, contamination prevention and efficient use of land resources. Some of the measures taken included:

- Trouble shooting and capital repair of pipeline transport.
- Modernisation and construction of facilities enhancing the petroleum gas utilisation level.
- Purchase of equipment for possible emergency oil and petroleum product spill response.
- Reclamation of impaired and oil contaminated lands.
- Integrated monitoring of environmental components
- Construction of new and reconstruction of existing water supply and water drainage networks.
- Assembly of chlorination units for the disinfection of industrial and river water.
- Replacement of equipment and optimisation of technologies aimed at lower pollutant emissions.
- Construction of a complex for oil-contaminated waste treatment.

A compliance audit of the LUKOIL Group HSE Management System was conducted in the reporting year. The auditors of Bureau Veritas, an international certification company, visited

21 organisations of the group. Following up on the results of the compliance audit of the LUKOIL Group organisations, 2 cases of material inconsistency in ZAO LUKOIL-AIK (eliminated in the course of the audit) and 17 cases of minor inconsistency were registered, which, however, require the development of corrective and preventive measures.

LUKOIL pursues active development of the Refining & Sales Business Segment, focusing its efforts on an increase in its oil conversion rate and the output of engine fuel with improved environmental characteristics.

The 2012-2021 strategy provides for:

- Considerable improvement of light petroleum products output at Russian refineries.
- A 70 per cent rise in gasoline manufactured at Russian refineries.
- Commissioning of new secondary oil-refining units at Russian refineries
- Bringing overseas refineries in line with the highest world standards.

Renewable Power Generation: The company recognises the huge resource and energy-saving potential of alternative energy sources. The priority types of alternative renewable energy sources encompass hydraulic power, wind power, solar power and geothermal power engineering. Today, the alternative energy forms are developed through cooperation with international companies focusing on renewable power generation. In 2011, LUKOIL Group promoted cooperation in the area of renewable power generation with Italy's ERG Renew that has vast experience in the construction and operation of renewable power generation facilities. The LUKOIL-ERG Renew partnership focuses on wind power projects in Bulgaria, Romania, Ukraine and Russia. Primarily, these are the projects of 'network' power stations in the countries of the company's active presence where the state provides administrative and economic support of the renewable power industry, including the projects of the 'network' wind power stations in Bulgaria and Romania.

LUKOIL is participating in the construction of a 100 megawatts (MW) solar power station in Uzbekistan jointly with the Uzbek Government and the Asian Development Bank. The end of 2011 saw the commissioning of a 1.25 MW photovoltaic power plant (PPP) at the LUKOIL Group TPP in Bulgaria. The expected yearly electric power generation volume is around 1,500 MW/h. The PPP operation will considerably reduce the negative environmental impact.

The company's achievements in the area of environmental safety were highly praised; LUKOIL was the winner of the 2011 contest 'National Ecological Award' in the category 'Innovative Eco-efficient Technologies in Industry.' In addition, LUKOIL won the Second All-Russian Contest for Healthy Lifestyle Projects, 'Healthy Russia,' in 'The Best Corporate Project' category held under the auspices of the RF Ministry of Healthcare and Social Development as part of the state programme for healthy lifestyle promotion.

Oil Pollution in Baltic Sea

In recent years, a number of new oil terminals have been built in the Baltic Sea area, resulting in increased transport of oil by ships and, consequently, an increased risk of accidents.

Transportation is a reason for 45 per cent of oil in the sea while an offshore production is a source of only 2 per cent of the oil pollution in the World Ocean. In the Baltic Sea, about 2,000 large ships and tankers are at sea every day. Thus, shipping activities, including oil transport and oil handled in harbours, have a major negative impact on the marine environment and coastal zone in the Baltic Sea. Oil and oily residue discharges from ships represent a significant threat to marine ecosystems. These discharges may occur during normal activities or may be accidental or illegal. The total annual number of oil spills into the Baltic Sea is estimated to be around 10,000 and the total amount of oil running into the sea can be as much as 10,000 tonnes which is considerably more than the amount of oil pouring into the sea in accidents. Oil spills cause the contamination of seawater, shores and beaches, which may persist for several months.

As far as oil exploitation at sea and on the coast is concerned, offshore operations are taking place for several years in Polish waters (two jackup rigs), until recently Germany operated two platforms very close to the coast. In March 2004, LUKOIL Kaliningradmorneft started to drill for oil and then to explore the Kravtsovskoe oil field from the D6 offshore oil platform in Russian waters. Latvia has plans to drill for oil in the water between Latvia and Lithuania (Marine Oil Pollution Information Gateway, 2004).

As far as the Baltic Sea ecosystem undergoes growing human-induced impacts, especially associated with an increasing oil transport, one of the main tasks in the ecological monitoring of the Baltic Sea is an operational satellite and aerial detection of oil spillages, determination of their characteristics, establishment of the pollution sources and forecast of probable trajectories of the oil spill transport.

Oil pollution monitoring in the Mediterranean, North and Baltic Sea is normally carried out by aircrafts, ships and satellites. Use of aircrafts and ships is expensive and is constrained by the limited availability of these resources. Aerial surveys over large areas of the seas to check for the presence of oil are limited to the daylight hours in good weather conditions. Satellite imagery can help greatly identifying probable spills over very large areas and then guiding aerial surveys for precise observation of specific locations. The Synthetic Aperture Radar (SAR) instrument, which can collect data independently of weather and light conditions, is an excellent tool to monitor and detect oil on water surfaces. This type of instrument is currently on board of the European Space Agency's (ESA) ENVISAT and ERS-2 satellites and the Canadian Space Agency's RADARSAT satellite.

Since 1993, there is no regular aerial surveillance of the oil spills in the Russian sector of the southeastern Baltic Sea. In early 2003, LUKOIL Kaliningradmorneft initiated a pilot project, aimed to the complex monitoring of the southeastern Baltic Sea, in connection with a beginning of oil production at continental shelf of Russia in March 2004. The large oil field 'Kravtsovskoe' (D6), discovered in 1983 in the southeastern Baltic Sea, is located at a distance of 22.5 km from the Kaliningrad region coast. A programme of ecological monitoring, which includes ship, coastal and satellite observations, was elaborated. Since 2003, regular environmental monitoring of the oil field is carried by R/V Professor Shtokman and ships of AtlantNIRO. According to the HELCOM recommendations, the monitoring net includes 22 regional oceanographic stations and 12 local stations located at distances of 100, 500 and 1,000 metres from the platform. Data received before the exploitation of the oil field provide initial background conditions of the marine environment. Today, measurements are used for the control of hydrological, hydrochemical and biological parameters of seawater from the surface to the bottom.

Operational satellite monitoring of the southeastern Baltic Sea, as an important component of the environment monitoring, was organised in June 2004 on the base of daily satellite remote sensing (ASAR [Advanced Synthetic Aperture Radar] ENVISAT, AVHRR NOAA, MODIS, QuikSCAT, JASON1). As a result, a complex information on oil pollution of the sea, sea surface temperature (SST), distribution of suspended matter, chlorophyll concentration, sea currents and meteorological parameters has been received. General goals of satellite oil pollution monitoring in the southeastern Baltic Sea are:

- Detection of oil spills in the vicinity of D6 oil platform (Kravtsovskoe oil field) as well as in the large area of the southeastern Baltic Sea.
- Identification of sources of pollution.

- Forecast of oil spills drift.
- Data systematisation and archiving.
- Cooperation with authorities.

Operational Oil Spill Monitoring by ASAR ENVISAT

The ENVISAT satellite was launched in March 2002 by the ESA. Operational systems, which include 10 instruments, have been developed to monitor oceans, ice, land and atmosphere. ENVISAT has 35 days repeat cycle, but due to wide swaths by some of the instruments, the Earth is covered within a few days. ASAR instrument is used for mapping sea ice and oil slick monitoring, measurements of ocean surface features (currents, fronts, eddies, internal waves), ship detection, oil and gas exploration, etc.

When oil is forming a thin layer on the sea surface it will damp the capillary waves. Due to the difference in backscatter signals from the surface covered by oil and areas without oil, radar satellites may detect oil spill slicks at the sea surface. SAR sensors have an advantage over airborne optical sensors in that they can provide data over large area under cloudy conditions and during darkness. Users of remotely sensed data for oil spill applications include the Coast Guard, national environmental protection agencies and departments, oil companies, ship ASAR ENVISAT coverage of the southeastern Baltic Sea (35 days cycle of descending and ascending passes), fisheries and oceans, and other organisations (ESA). Ground stations, such as KSAT Station (Kongsberg Satellite Services, Tromso, Norway), acquire ASAR ENVISAT data and distribute them to customers buying data for operational use in the near real-time. Since July 2004, LUKOIL Kaliningradmorneft via Sea Venture Bureau, which signed such a contract with KSAT, during six months was receiving ASAR data from all ENVISAT passes over the southeastern Baltic Sea in operational regime (12 hours after the satellite's overpass) for further expert analysis and detection of oil spills. In total, 86 oil spills have been detected in 88 ASAR ENVISAT images (400 x 400 km, 75 m/pixel resolution) received in June December 2004. Oil spill gallery is presented in the report.

For interpretation of ASAR ENVISAT imagery and forecast of oil spills drift, IR and VIS AVHRR (NOAA) and MODIS (Terra and Aqua) images are received, processed and analysed, as well as the QuikSCAT scatterometer and the JASON1 altimeter data. The results obtained in 2004 show the efficiency of this approach to the complex satellite monitoring of the ecological state in the southeastern Baltic Sea. Satellite receiving station at Marine Hydrophysical Institute (MHI) in Sevastopol is used for operational 24 hours/day, 7 days/week receiving, processing and analysis of AVHRR NOAA data for construction of sea surface temperature, optical characteristics of sea water and currents maps. The Baltic Sea Satellite Monitoring Center holds special internet server in MHI as a platform for data collection, archiving and exchange between participants of the project via File Tranfer Protocol (FTP), as satellites images typically are too large to be transferred by email.

Reconstruction of the Currents Field

Sequence of daily MODIS IR and VIS imagery allows to reconstruct a real field of surface currents (direction and velocity of currents) with 0.251 km resolution which is very important for a forecast of a direction and velocity of potential pollution drift including oil spills. Sea wind speed fields are derived from scatterometer data from every path of the satellite over the Baltic Sea (twice a day). These data are combined with data from coastal meteorological stations in Russia, Lithuania, Latvia, Estonia, Sweden, Denmark, Germany and Poland for the analysis of ASAR ENVISAT imagery.

Wave Height by JASON1 Satellite Altimetry Data and Simulation Results

Satellite altimetry data from every track over the Baltic Sea are used for compilation of sea wave height charts, which include the results of the Fleet Numerical Meteorology and Oceanography Center (FNMOC) WW3 Model. Exact repeat cycle of along track measurements is 10 days. The interactive numerical model Seatrack Web SMHI (The Swedish Meteorological and Hydrological Institute) is used for a forecast of the drift of: (1) all large oil spills detected by ASAR ENVISAT in the southeastern Baltic Sea and (2) virtual (simulated) oil spills from the D6 platform. The latter is done daily for two purposes: operational correction of the action plan for accident elimination at the D6 and ecological risk assessment (oil pollution of the sea and the Curonian Spit).

This version of a numerical model on the Internet platform has been developed at SMHI in close cooperation with Danish authorities. The system is based on an operational weather model HIRLAM (High Resolution Limited Area Model, 22 km grid) and circulation model HIROMB (High Resolution Operational Model for the Baltic Sea, 24 layers), which calculates the current field at three nautical miles grid. The model allows to forecast the oil drift for two days ahead or to make a hind cast (backward calculation) for eight days in the whole Baltic Sea. When calculating the oil drift, wind and current forecasts are taken from the opertional models. An oil spreading calculation is added to the currents. This powerful system today is in operational use in Sweden, Denmark, Finland, Poland, Estonia, Latvia, Lithuania and Russia. Statistics, based on daily forecast of the oil spills drift from the D6 oil platform in July-December 2004, shows potential probability (%) of the appearance of an oil spill in any point of the area or direction of the oil spill drift during 48 hours after an accidental release of 10 m^3 oil. Probability of the oil spill drift directed to the Curonian Spit (150°sector in red) equals to 67 per cent, but only in a half of these cases, 35 per cent oil spills reached the coast.

Case of Petrobras

Petróleo Brasileiro SA or Petrobras (Petrobras) is a Brazil-based integrated oil and gas company. The company divides its activities into six segments: Exploration and Production; Provision; Gas and Energy; Biofuel; Distribution; and International. Directly or through its subsidiaries, Petrobras is engaged in the research, extraction, refining, processing, trade and transport of oil from wells, shale and other rocks, its derivatives, natural gas and other liquid hydrocarbons, as well as in activities related to energy, development, production, transport, distribution and commercialisation of energy. As of 31 December 2011, the company owned 132 concession contracts for 194 exploration blocks, and had a number of subsidiaries, including Petrobras Quimica SA, Petrobras Distribuidora SA, Braspetro Oil Services Company and Braspetro Oil Company, among others.

Petrobras engages in oil and gas activities. It operates through the following segments:

- Exploration and Production (Exploration and Development).
- Refining, Transportation and Marketing (Supply).
- Distribution, and Natural Gas and Power (Gas and Energy).

The Exploration and Development segment include exploration, development and production activities in Brazil. The Supply segment encompasses refining, logistics, transportation, and the purchase of crude oil, as well as the purchase and sale of oil products and fuel alcohol. It also include the petrochemical and fertilisers division, which includes investments in domestic petrochemical companies and the company's two domestic fertiliser plants. The Distribution segment encompasses oil product and fuel alcohol distribution activities conducted by the company's subsidiary, Petrobras Distribuidora S.A. The Gas and Energy segment encompasses the purchase, sale and transportation of natural gas produced in or imported into Brazil.

Petrobras is the third largest energy company in the world, with an average daily output of 2.6 million boe. Products made by the Petrobras System—formed by the company, its subsidiaries, jointly controlled and associated companies—are present in virtually all areas of modern life, ranging from the fuel that moves cars and propels aircraft to the liquefied petroleum gas (LPG) used in the kitchens of homes and businesses. Diesel fuel, gasoline, lubricants, naphtha, biodiesel, ethanol, fuel oil and jet fuel, among others, complement the company's portfolio.

Activities: From the well to the service station and much more. Petrobras proudly states that it generates the energy that sets the country into motion. It also performs well in several areas such as research, innovation and beside providing with quality products.

It has expanded its operations aiming to be among the top five integrated energy companies in the world by 2020. The 2012-2016 Business Plan calls for investments in the order of $236.5 billion.

Corporate Strategy

Integrated growth, profitability and socio-environmental responsibility are keywords in its corporate strategy. It is based on the performance of three pillars that built the 2020 Mission and Vision, transparently and with attentive eyes on what is going on in Brazil and in the world. Operate in a safe and profitable manner in Brazil and abroad, with social and environmental responsibility, providing products and services that meet clients' needs and that contribute to the development of Brazil and the countries in which it operates.

Vision 2020 aims to be one of the five largest integrated energy companies in the world and the preferred choice among the stakeholders. Described in the Strategic Plan, the values are the ways the company guides its strategies, actions and projects. They must appear in how business is carried out and reflect Petrobras' way of being.

It believes in pursuing business success under a long-term perspective, contributing to economic and social development and to a healthy environment in the communities in which it operates. It seeks to maximise collaboration and to seize synergies among teams, areas and units, ensuring an integrated vision of the company in its actions and decisions; and to continuously to generate value for all stakeholders, focusing on capital discipline and on cost management. It values and acknowledges, in a differentiated manner, high-performance people and teams. One important dimension of Petrobras is readiness for change and also accepting the responsibility for inspiring and creating positive change.

Entrepreneurship and innovation: Petrobras cultivates the overcoming of challenges and seeks, relentlessly, to generate and implement innovative technological and business solutions that help Petrobras attain its strategic objectives.

Ethics and transparency: Its business, actions, commitments and other relationships are guided by Petrobras System's Ethical Principles.

Respect for life: Petrobras respects life in all of its forms, expressions and situations, and seeks excellence in matters that involve health, safety and the environment.

Human and cultural diversity: Petrobras values human and cultural diversity in our relationships with people and institutions. The company ensures the principles of respect for difference, non-discrimination and of equal opportunities.

People: Petrobras turns people and their development into a performance differential of Petrobras.

Proud to be Petrobras: Petrobras is proud to belong to a Brazilian company that makes the difference no matter where it operates, for its history, achievements and for its capacity to overcome challenges.

Environment Management

Integrated Actions

Petrobras seeks to minimise the impact of its operations and products on the environment. This concern is featured both in the company's Business Plan and in its Corporate Strategy. Managing potential environmental risks inherent to the oil and gas industry such as the consumption of natural resources, air emissions, interference in the territories, impacts on biodiversity and pollution caused by waste requires integrated actions in the environmental area and should involve all Petrobras System areas, units and subsidiaries, from the strategic to the operational level.

The governance model established by Petrobras contributes to the integration of company actions and initiatives in the environmental area and calls for an Environment Committee, which is linked to the Board of Directors, a Business Committee and four Integration Committees. The Committee for the Integration of Technology, Engineering and Services brings together executives from various company areas, and its agenda includes discussing and evaluating Petrobras' environmental management and performance. An additional discussion forum is promoted by commissions connected to the Committee. Consisting of safety, health and environment managers, as well as energy efficiency managers from the business areas, services and subsidiaries, these commissions allow a deeper understanding of and engagement with safety, health and environment management issues, as well as environmental licensing and compensation, energy efficiency, emissions and climate change.

In 2010, Petrobras' expenditures with the environment surpassed \$2.4 billion, about \$457 million more than in 2009. At Petrobras, environmental management is also integrated to the operational safety, energy efficiency and health management given the synergy between all of these issues. This integrated management model is identified by the acronym SEEH (safety, environment, energy efficiency and health).

There are 15 corporate guidelines that set out the requirements to be met by the HSE management system. Standards derived from these guidelines are applied in all of Petrobras' activities in Brazil and abroad. The adherence of the units' management systems to these standards is checked regularly by internal audits. The integrated management process has ensured significant improvements in Petrobras' HSE performance, which currently approaches—and in some cases exceeds—global oil and gas industry benchmarks of excellence.

Shared Responsibility

The corporate area in charge of guiding, monitoring, evaluating and communicating HSE management at Petrobras is the SEEH management. All leadership levels, however, are responsible for implementing and using the management system and, ultimately, for their area or unit's HSE performance. The HSE management system's governance is ensured by a structure of committees and commissions that involve all of the company's management levels.

Petrobras is also investing in new technologies and in more environment friendly energy sources, in increasing process efficiency, reducing natural resource consumption and emissions, and in ecosystem conservation and restoration.

The company seeks to encourage its suppliers to also adopt environmentally responsible attitudes and practices through requirements imposed in the supplier corporate register, contract clauses, by means of the structured system for handling non-compliances and, ultimately, by applying sanctions that may prevent the companies from providing their products or services to Petrobras. In late 2010, at least 10 companies were in this situation.

There are no significant gaps in the management process of HSE-related aspects connected to Petrobras' activities. The company's actions in this area relate specifically to the pursuit for continuous performance improvement, to the identification of new demands and expectations of our stakeholders on socioenvironmental issues, and to transparency in the relationship with these groups. This focus on the stakeholders has led Petrobras to emphasise a few environmental issues.

Emissions Reductions and Management

Climate Change

The Brazilian economic growth over the upcoming years may increase the energy sector GHG emissions. Given this perspective, the company included Strategic Climate Change and Energy Efficiency Projects in its Strategic Plan. These projects were designed to help Petrobras voluntarily reach levels of excellence in the oil and gas industry with regard both to processes and products intensity of GHG emission and to processes relating to energy efficiency.

Petrobras set the following targets to be achieved by 2015:

- Reduce the energy intensity of refining and thermoelectric plant operations by 10 per cent and 5 per cent, respectively.
- Reduce natural gas flaring intensity in exploration and production operations by 65 per cent.
- Reduce the intensity of GHG emission from exploration and production, refining, and thermoelectric plants operations by 15 per cent, 8 per cent and 5 per cent respectively.

Petrobras also considers it essential to deeply analyse both the potential impact and the possible worsening of weather phenomena may have on its operations, and also the effects of regulation and market changes targeting the mitigation of global climate change. Petrobras' operations in the US and in Japan are currently the ones that are most likely to have restrictions concerning GHG emissions, since these countries have already made commitments to mitigate global climate change.

Here are the major climate change-related initiatives Petrobras has adopted:

- Air Emissions Management System (Sigea), which, since 2002, has assisted in the preparation of detailed inventories of GHG emissions and other air pollutants, recording data from over 30,000 sources of emissions in all of the company's units in Brazil and abroad.

- Establishment of an internal governance system that allows the company to assess opportunities to reduce the intensity of GHG emissions associated with new investment projects.
- Investments in biofuels: Established in 2008, Petrobras Biocombustível has the capacity to produce 498,000 cubic meters of biodiesel per year. The 2010-2014 Business Plan foresees investments of $3.5 billion in biofuels, of which $530 million will be earmarked for research and development of new technology, including for second generation biofuels.
- Development of internal programmes to increase operational energy efficiency.
- Gas flaring reduction and use.
- Power generation from renewable sources.
- Encouraging rational fuel use through actions carried out under the National Program for Oil Products and Natural Gas Use Rationalization.
- Investments in R&D programmes focused on energy efficiency, renewable energy, and on CCGS (carbon capture and geological storage).
- Engagement with organisations in Brazil and abroad that are involved in efforts to mitigate global climate change, and adaptation to these changes.
- Sponsorship for environmental projects of water conservation, carbon fixation, and emissions mitigation by recovering degraded areas or using such areas to establish sustainable production systems and natural areas and forests conservation.

Debating Climate Change

Petrobras participates in discussions about global climate change in several forums, with special emphasis on the 16th Conference of the Parties of UNFCCC[4] where the company presented details on

4. Executive Committee of the UN Framework Convention on Climate Change.

its emissions monitoring system—the Air Emissions Management System (Sigea). Petrobras also takes part in other initiatives, such as the IPEA[5] Climate Change Working Group (CCWG), the Arpel Climate Change and Energy Efficiency Committee,[6] World Business Council for Sustainable Development (WBCSD) initiatives in the area, and in the Thematic Chamber for Energy and Climate Change of the CEBDS,[7] the Brazilian local network of the WBCSD.

Risks and Opportunities

In its planning processes, Petrobras analyses the potential impacts climate change may have on its business. Based on the 2030 Corporate Scenarios, the company seeks to predict the macro-trends related to these issues. The scenarios involve aspects regarding production processes and energy consumption habits, international negotiations, nations climate and energy policies in force and their effects on the carbon market.

International agreements and national and regional legislation and regulatory measures designed to minimise GHG emissions are in different stages of discussion and implementation around the world. The National Policy on Climate Change was regulated in December 2010. As a result, Brazil voluntarily set a limit for its GHG emissions. The imposition of fiscal, taxation or other measures to discourage fossil fuel use or to promote low-carbon technologies may have negative impacts on the oil market, but can also create market opportunities for renewable fuels.

In the technology area, Petrobras' R&D Centre (Cenpes) also assesses the risks associated with climate change, partnering with universities, government institutions and other research centres.

5. International Petroleum Industry Environmental Conservation Association.

6. Regional Association of Oil and Natural Gas Companies in Latin America and the Caribbean.

7. Brazilian Business Council for Sustainable Development. In 2010, Petrobras was awarded in the 2010 Carbon Leadership Awards, organised by British magazine The New Economy, in the "Best Emissions Reporting" category. The award recognises companies that stand out both in reducing and in disclosing their carbon emissions.

It also contributes to the development and operation of the Thematic Network on Climate Change, which is focused on technical cooperation and financial support for science and technology organisations nationwide. Created in 2008 by the National Institute for Space Research and the Ministry of Science and Technology, the network comprises 12 institutions and aims to develop national capacity in the carbon capture, transport and storage areas. Between 2006 and 2009, the company invested $30 million in climate change and carbon sequestration research technologies. Additional investments of $200 million are expected for the 2010-2015 period.

The R&D programmes to enable the capture, sequestration, transport and geological storage of carbon dioxide are: Technological Program for Climate Change (Proclima), created in 2007, and Technological Program for CO_2 Management for the Pre-Salt Development (PRO-CO_2), created in 2009, with a medium-term perspective and focused on CO_2 emissions in the development of the Santos Basin Pre-Salt Layer.

Since the 1970s, Petrobras has operated in several stages of the ethanol value chain through the National Ethanol Program (Proálcool). Additionally, it created the Biofuels Technology Program (technological development for first- and second-generation biofuels), which is expected to receive $530 million in investments between 2010 and 2014.

Extreme weather events, which can be worsened by global warming, may potentially affect:

- Global operations in deep waters, which can be affected by sea level rise and by more intense and frequent hurricanes and storms.
- Road, maritime, inland waterway and pipeline transport activities, which carry large volumes of oil, oil products, gas and biofuels.
- The availability of water, an essential resource for many of the company's operations.

- Biofuel production, in the event of a change in rainfall patterns that could affect the production of the agricultural inputs that are used as feedstock for these products.

Extreme weather events may cause facility insurance premiums to surge and demand additional investments in order to ensure the security and integrity of such facilities.

Energy Efficiency and Emission Management

The strategic objective of maximising energy efficiency and reducing the intensity of GHG emission requires the company to maintain a portfolio of projects, which will receive $976 million in resources in the period ranging from 2010 to 2015.

Investment-Emission Reduction Ratio

Between 2006 and 2010, the company invested $404 million in projects directly related to energy efficiency, affording savings of up to 2,740 barrels of oil equivalent per day (boed).

- Does not include gas flaring or the gas used as input hydrogen generation at refineries and fertiliser plants; includes fuels used to generate electricity at the thermoelectric plants (natural gas, diesel and fuel oil); does not include Petrobras Biocombustível data. Due to changes made to the fuel consumption analysis criteria, information on 2008 and 2009 was updated.
- In terajoules: Due to both the increased dispatch of Petrobras' thermoelectric plants (TEP) and the consumption of energy by the other activities, which kept pace with company's growth, there was an 18.6 per cent rise in the company's total energy consumption in 2010, 112,602 terajoules more than in 2009.

Petrobras has the Internal Energy Efficiency Program, whose actions are monitored by the Commission for Energy Efficiency, Air Emissions and Climate Change in conjunction with the Internal Commissions for Energy Conservation (Cices).

The energy efficiency initiatives are associated with:

- Energy management.
- Energy optimisation and integration.
- Cogeneration.
- Turboexpanders.
- Advanced process control.
- Facility modernisation.
- Equipment adaptation.
- Reduction of gas flaring.
- Reduction of steam, condensate and water losses.
- Standardisation of designs and operating systems.

Development/Innovation

The role played by R&D in the pursuit of excellence in energy efficiency, GHG emission management and in global climate change mitigation is worthy of note. The expected investments in these areas, approved under the R&D Business Plan, amount to $200 million for the 2010-2015 period, with emphasis on the Technological Program for Climate Change (Proclima) and on the Technological Program for CO_2 Management (PRO-CO_2), focusing on projects intended for CO_2 capture and geological storage, energy efficiency and other mitigation technologies.

Examples of energy efficiency R&D projects include the development of real-time process optimisation tools, more efficient combustion and heat transfer equipment, and assessment methodologies and tools to diagnose and identify energy efficiency improvement opportunities.

The company also takes part in the Ministry of Mines and Energy's National Program for Oil Products and Natural Gas Use Rationalization (Conpet), which in 2010 afforded savings of 72.3 million litres of diesel through the Economizar e Transportar (Save and Transport) projects, and of 524,000 cubic metres of gas in the residential sector by implementing energy efficiency measures and equipment labelling. These results kept upwards of 1 million tonnes of CO_2 equivalent and of 4,000 tonnes of particulate matter from being released into the atmosphere.

Alternative and Renewable Energy

Initiatives geared to encourage the use of energy generated by renewable resources include biofuel production and marketing. In 2010, the consumption of biodiesel produced by Petrobras Biocombustível prevented nearly 1 million tonnes of CO_2 emissions. New projects were initiated aimed at producing biodiesel from palm oil in the state of Pará to meet the demand both in Northern Brazil and biodiesel production projects in Portugal.

The strategy in the ethanol segment is directed towards acquiring stakes and signing on partnerships with companies operating in the industry, guaranteeing technological excellence in sustainable biofuel production. In consequence, the subsidiary now has interests in 10 ethanol plants with a capacity to crush more than 24 million tonnes of sugar cane and to produce in excess of 900 million litres of ethanol per year, which could potentially prevent the emissions, by customers and consumers, of about 1.4 million tonnes of CO_2, arising from avoided gasoline consumption.

In 2009, Petrobras started converting the Juiz de Fora Thermoelectric Plant to run on ethanol and on natural gas. This could prevent about 100,000 tonnes of CO_2 emissions per year. Using ethanol in a gas-fired turbine is unprecedented in the world.

An important initiative is being implemented at the Duque de Caxias Refinery (Reduc) to use biogas generated at the Gramacho Landfill as a source of energy. Gas supply is planned to start in May 2011. It is estimated that 160 million cubic metres of biogas will be consumed per year, avoiding the emission of 75 million cubic metres of methane into the atmosphere.

Emissions Reductions and Management

GHG Emissions

From 2006 to 2010, Petrobras prevented the emissions of approximately 6 million tonnes of CO_2 equivalent from its operations. This was achieved primarily by reducing flaring, by applying new technologies and by using renewable energy.

In 2010, Petrobras' operations released 61 million tonnes of CO_2 equivalent into the atmosphere. The higher direct emissions figure, compared to the previous year, was due to the inclusion of data regarding the Pasadena Refinery, in the US, and to the increased dispatch of electricity by thermoelectric plants; in contrast, there was significant reduction of gas flaring in the exploration and production (E&P) area.

Petrobras discloses its GHG emissions inventory voluntarily, both in its own reports and by taking part in initiatives such as the Brazil GHG Protocol, of which it is a founding member, and the Carbon Disclosure Project (CDP). It also collaborates with Federal Government and state government actions in this area. The company participated, for example, in the preparation of the Second National Communication on Greenhouse Gas Emissions, the Brazilian emission inventory, and conducted an inventory of its GHG emissions in the state of São Paulo aiming to assist in the State Policy on Climate Change.

Results of the E&P, refining, fertiliser, petrochemical, power generation, ground (pipeline and road) and maritime transportation operations, as well as distribution activities in Brazil, Argentina, Bolivia, Colombia, Ecuador, Mexico, Peru, Paraguay, Uruguay and the US are as follows:

- Indirect emissions refer to the purchase of electricity and steam, supplied by third parties, in the mentioned countries. The air emissions inventory is prepared according to the guidelines of the GHG Protocol, a corporate standard (World

Research Institure [WRI]/WBCSD). The scope of the inventory is the operational control and/or financial control. As such, it includes assets of which Petrobras is the operator and assets in which it holds 50 per cent or more of the financial control. The approach used to prepare the inventory is based on the bottom-up methodology, i.e. the total inventory is the sum of the emissions from each monitored source. The algorithms used to calculate GHG emissions are based on publicly accessible international benchmarks, such as the API Compendium (API) and the 'AP-42' (USEPA)]. Emissions are calculated based on the input of specific data that consider the characteristics of the equipment, processes and fuel consumption.

All algorithms are executed in a computerised system, the Sigea, which allows for the monthly monitoring of the inventory of approximately 30,000 emission sources. In terms of GHG emissions, the inventory includes emissions of CO_2, CH_4 (methane) and N_2O (nitrous oxide). The results are shown in millions of metric tonnes of CO_2 equivalent (CO_2e), calculated according to the *Second Assessment Report* of the Intergovernmental Panel on Climate Change (IPCC). Petrobras submits its inventories, periodically, to a verification process carried out by the third party pursuant to the ISO 14064 standard. As a result, inventories from 2002 to 2007 were verified in two cycles. The first included the period ranging from 2002 to 2004, while the second from 2005 to 2007. The 2009 direct emissions were revised. The result of 62 million tonnes of CO_2e published in the *2009 Sustainability Report* was revised during the critical assessment process carried out for the results. Other indirect GHG emissions had not been consolidated yet when this report was closed.

Carbon Credits

Two projects developed voluntarily by Petrobras were registered by the Executive Committee of the UNFCCC as compliant with the requirements of the Kyoto Protocol Clean Development Mechanism, and, thus, are candidates for carbon offset crediting. The first, registered in March 2007, was the Macau wind farm, which is installed in Rio Grande do Norte and is capable of producing 1.8 MW of electricity and of avoiding the emissions of approximately 1,300 tonnes of CO_2 equivalent per year.

The second, registered in October 2009, was the implementation of a project to offset N_2O emissions at Petrobras' nitrogen fertiliser plant in Bahia. This initiative prevents the release of approximately 57,000 tonnes of CO_2e per year into the atmosphere.

Pre-salt

During 2010, media and websites reported that CO_2 rates of Pre-salt wells were above average and that could cause more emissions. However, tests show that the concentrations vary: in some wells, they are above the Santos Basin average; in others, close to zero. Although there are no specific limits on carbon dioxide emissions in oil production in Brazil, Petrobras is committed to preventing the release of CO_2 coming from the Pre-Salt region into the atmosphere. To achieve this, new technologies designed to capture, transport and geologically store the CO_2 associated with oil and gas flaring will be developed.

The main measure that is being studied is underground reinjection, a process that is already carried out in countries such as Norway. Other options include CO_2 storage in reservoirs that have saline aquifers nestled under the seabed, reinjection it in depleted gas reservoirs, and storing CO_2 in caves in the salt layer. The likely emissions originating from Pre-salt exploration are featured in the Ten Year Plan for Energy Expansion (PDE 2019), prepared by the Energy Research Corporation, an agency of the Brazilian Government.

Operations in the Pre-salt

It all started with an idea of those who believed that there are no barriers, rather opportunities. Petrobras have an even better horizon ahead. The discoveries made in the pre-salt raised Petrobras to a new level of reserves and oil production, ranking it in a prominent position among the major energy companies.

With the experience they have acquired developing fields nestled in deep waters, our technicians are now ready to develop the accumulations discovered in the pre-salt. To this end, they are already adapting the technology and the logistics the company developed through the years. The goal is to achieve, by 2017, a daily production in excess of a million barrels of oil in the Pre-salt areas it operates.

Case of Ecopetrol S.A.

Ecopetrol S.A. is a Colombia-based company engaged in the development of commercial or industrial activities related to the exploration, exploitation, refining, transportation, storage and distribution of hydrocarbons and its by-products. The company divides its operations into four business segments: Exploration and Production, Transportation, Refining, and Marketing and Supply. The Company's range of products include such commodities as middle distillates, gasoline, natural gas, LPG, propane, asphalts and crude oils, among others. As of 31 December 2011, the Company owned such subsidiaries as Ecopetrol Oleo e Gas do Brasil Ltda, Ecopetrol del Peru SA, Black Gold Re Ltd, Andean Chemicals Ltd, Bioenergy SA, Refineria de Cartagena SA, Hocol Petroleum Limited and Oleoducto Bicentenario de Colombia SAS, among others.

Ecopetrol SA is among the top 50 oil & gas companies in the world and the fourth-largest in Latin America. In terms of revenue, profit, earnings before interest, taxes, depreciation and amortization (EBITDA), assets and equity, Ecopetrol SA ranks first in Colombia and is the country's only vertically integrated oil & gas company. Ecopetrol has presence in Colombia, Brazil, Peru and the US (Gulf Coast). Ecopetrol SA operates in the following business segments: (1) exploration and production; (2) transportation and logistics; (3)

refining, petrochemicals and biofuels; and (4) sales and marketing of crude oil, natural gas and refined products. Ecopetrol holds stakes in the following companies:

- E&P: Hocol (100%), Savia (50%), Equion (51%), Ecopetrol America (100%), Ecopetrol Peru (100%), Ecopetrol Brasil (100%).
- Transportation: ODL (65%), ODC (73%), OBC (56%), Ocensa (72%) and CENIT (100%).
- Refining and Petrochemical: Reficar (100%) and Propilco (100%).
- Biofuels: Ecodiesel (50%) and Bioenergy (88%).

Ecopetrol's shares are listed on the Colombian Stock Exchange (ECOPETROL) and its ADR's (American Depository Receipt) are listed on the New York Stock Exchange (EC) and the Toronto Stock Exchange (ECP). The Republic of Colombia is the major shareholder of Ecopetrol, with an interest of 88.5 per cent.

Strategic Plan

Ecopetrol is carrying out its strategic growth plan for the year 2020 with a capex of approximately $80 billion. The main goals for 2020 are: (1) producing 1.3 million boed (including contributions from affiliates and subsidiaries), (2) increasing average 1P reserve life to approximately 10 years by adding 6,200 million barrels, (3) modernising existing refineries and increasing their capacity from 330 to 415 thousand processed barrels/day with a conversion factor of 98 per cent, (4) reaching a production of 450,000 tonnes per year of biofuels, and (5) increasing transport capacity to 1.7 million barrels/day.

Corporate Governance Practices in Ecopetrol

With the intention to generate confidence among stakeholders and ensure business sustainability, Ecopetrol adopted transparency, governance and control practices to ensure good corporate governance. Following sub-sections present the corporate governance model and the mechanisms used by Ecopetrol to address the application of best practices.

Environmental Dimension

This dimension includes the initiatives developed by the company to reduce its environmental impact towards the sustainable development of the country.

In 2011, Ecopetrol invested US $1,3,11,635 million in environmental programmes. This value represents an increase of 33 per cent compared to 2011.

In addition to environmental legal compliance, Ecopetrol has defined three strategic pillars for environmental management:

(i) Biodiversity: Ecopetrol aims to operate while providing sustainability to the ecological systems in which they operate. This will be achieved through joint work with stakeholders for the promotion of projects on the conservation, management and use of biodiversity.

(ii) Eco-efficiency: It aims to progressively reduce the environmental impacts of the operation and the investment projects. Specifically, in the reduction of discharges and solid waste.

(iii) Climate change: It aims to operate in line with the UNFCCC. Towards this, much focus will be on reducing emissions of GHGs and ensuring at least four new energy sources.

Action Framework

Between 2007 and 2009, Ecopetrol redefined its environmental strategy to go beyond what is required by law and become a world-class company, recognised, among other things, for the commitment to environmental care. The environmental issue was made visible in the Strategic Framework, emphasising that it is a corporate responsibility that explicitly establishes the recognition of the value of life and natural resources before starting any project. Therefore, Ecopetrol aims to operate in harmony and balance with the stakeholders and the environment.

Biodiversity

Biodiversity is identified as one of the strategic areas for Ecopetrol's environmental management. Here, the company found a new challenge: managing the biodiversity of its areas of influence despite the environmental background in the regions and permanent changes resulting from the dynamics of a company like Ecopetrol.

The direction has always been very clear: to work with the highest national and regional authorities to manage biodiversity, and seek to strengthen regional and local actors that drive processes of change in the regions where we operate.

Adopt Your Logo

Adopte su Logois is an initiative from Instituto Alexander von Humboldt and Ecopetrol aims to generate strategies for the understanding and conservation of the animal species that represent their businesses and brands. The aim is that this initiative will be adopted by national entities that are also represented by logos containing wildlife, and for information to be generated that enables their management and conservation. Similarly these efforts will make it possible to support research processes on other 'orphan' species that require special attention due to their vulnerability. The project started in 2012 and its results will start being published in 2013.

National Biodiversity Call

The National Biodiversity Call is an initiative created with the aim of promoting the development of biodiversity research projects. It was first carried out in 2009 with support from the Ministry of Environment, Housing and Territorial Development (now the Ministry of Environment and Sustainable Development). In its first iteration, 10 projects were selected relating to the understanding, management and use of biodiversity in High Andean Wetlands and

Ecopetrol's areas of influence, and in its second iteration, which took place in 2011, 12 projects were selected to be undertaken in the same area of influence defined for 2009. The institutions that submitted the best projects and met the contest's conditions won up to 50 per cent of the project's funding, which could last from 8 to 24 months.

Regional Projects

Ecopetrol has undertaken projects that are related to the understanding, use and management of biodiversity at the regional level, within the areas of their influence and includes:

(i) American crocodile: As a part of biodiversity programme, in 2011, Ecopetrol has undertaken the implementation of a project to evaluate populations of Crocodylus Acutus (American crocodile) with the support of Ecopetrol, the Regional Autonomous Corporation of the Northeast Border (CORPONOR) and the Ministry of Environment. These species are found in Appendix I of the CITES (Convention on International Trade in Endangered Species of Wild Fauna and Flora), those which are in danger of extinction, threatened since the last century due to the unlawful use of their skin.

This study seeks to carry out projects for the assessment and preservation of species in strategic areas of operation, projects with species in danger of extinction, and projects for species of special interest to the communities of these areas. The study made it possible to scientifically document the American crocodile population in the rivers Sardinata, San Miguel, Nuevo Presidente and Tibú, from the Catatumbo basin, and found a natural state population of 1.48 animals per kilometrw, over a length of 132 kilometres, i.e. 196 crocodiles established in the basin. Also, the preliminary strategies were defined to ensure the conservation and sustainability of the species, with the help of the community. For this, among other points, it is specified that there is a need for further research, education, community involvement and a declaration of protected areas.

Climate Change

In mid-2011, Ecopetrol began the creation of its climate strategy (2012-2020), which seeks to guide, plan, define and coordinate the company´s efforts on the issue of climate change, as well as to track each of the proposed lines of action. It identified the following areas and lines of action following the discussions held between different departments of the company:

Clean Development Mechanisms (CDM): Within the framework of the agreement signed between Ecopetrol and the Inter-American Development Bank (IDB) in 2011, a CDM feasibility analysis was carried out for each of the energy components that form part of the expansion and modernisation of the Cartagena Refinery, determining the potential for reducing GHG emissions, and their applicability under the criteria defined by CDM, which made it possible to establish requirements compliance for each of them, initiating the preparation of the Project Design Document.

In order to access best business practices and technologies worldwide that make it possible to achieve better energy indicators and thus reduce GHG emissions, the company is working on the formalisation of technical cooperation agreements with international bodies such as the Global Methane Initiative (GMI) and the Government of Canada, which are worldwide leaders in the implementation of technology and the evaluation of projects for energy optimisation to reduce the carbon footprint of the crude oil produced, as well as for intermediate goods and associated petrochemicals.

Atmospheric emissions: In implementing the continuous improvement process for the emissions inventory, during 2011 the following activities were carried out:

- Pre-audit for inventory verification in accordance with the guidelines and requirements established by ISO 14064, which made it possible to identify opportunities and actions for improvement to the inventory, which will be carried out during 2012. Inclusion of new operations in the estimate made for 2011 and adjustment of some emission factors.

- GHG Emissions: In 2011, GHG emissions were estimated at 6,722 kilotonnes of CO_2e, an increase of 17 per cent compared to 2010, mainly due to the inclusion within the inventory of emissions from the Northeast Management Department, an area that became part of Ecopetrol's operations in June 2010.[8]

Eco-efficiency

In order to achieve the Mega target of producing 1.3 million clean barrels per day in 2020, the company has been making significant efforts in relation to eco-efficiency: better use of resources, rational use of energy sources, reduction in discharges, waste, emissions and costs associated with energy consumption and products.

Biofuels: Biofuels are fuels produced from biomass, i.e. organic matter that comes from a biological process, which is usable as an energy source. According to the raw material and technology used in its production, biofuels can be classified into first, second and third generation. First-generation biofuels use food grade raw materials and process technologies such as fermentation and transesterification; it is mainly these that have reached a commercial production stage. Second-generation biofuels are derived from raw materials that have no food use, through the aforementioned processes and through thermochemical processes. Finally, third-generation biofuels use similar production methods in specifically designed or adapted bioenergy crops; these are still at an early stage of development and are a long way from being produced on a commercial scale.

Some noteworthy aspects of biofuels are:

- Their raw material is 100 per cent renewable.
- They reduce total emissions of GHGs.

8. The above figures corresponds to information inventoried for the first three quarters and averaged for the fourth quarter. However, this data may change once the definitive information is obtained. *Source:* Ecopetrol. Instituto Colombiano del Petróleo. Vice Presidency of Services and Technology.

- They can be used in a certain mixture percentage with fossil fuels without requiring modifications to vehicle engines.

Colombia entered the biofuel sector during the previous decade, defining the regulatory framework for the production and commercialisation of ethanol and biodiesel within the country, in 2001 and 2004 respectively, and starting mixtures thereof with fossil fuels in 2005 and 2008, respectively. The use of biofuels within the country is a consequence of the aim to diversify the energy mix, with criteria such as: employment generation, agro-industrial development, environmental sustainability and improvement of fuel quality.

Ecopetrol took its first steps within the sector in 2007, and now, biofuels are an integral part of the strategic framework as one of its guiding principles. They produce and market biofuels through Ecopetrol Corporate Group companies and are currently producing and marketing palm biodiesel in Ecodiesel Colombia S.A., as well as developing a project to produce ethanol from sugar cane in Bioenergy S.A.

Strategy: The strategic guiding principle for Biofuels is focused on the domestic market. The mega target for biofuels consists in the production of 450,000 tonnes in 2020. The biofuels strategy focuses on:

- Profitable consolidation in the local market.
- Competitiveness in the biofuels value chain.
- Development of growth opportunities in the domestic market.
- Our projects.

In 2007, the company Ecodiesel Colombia S.A. was established and the project began for the construction of its production plant for biodiesel from palm oil, which began commercial operation in 2010. Moreover, in 2008, Ecopetrol gained shareholding in the company Bioenergy S.A. in order to grow and execute a project to produce ethanol from sugar cane in the department of Meta.

Energy management: In order to achieve the mega target of reaching 1,300,000 clean barrels, the company has been making significant efforts in energy, increasing the efficient and rational use of energy sources used in the organisation's production processes and seeking to reduce consumption, costs, CO_2 emissions and risk in the use of sources, thereby increasing operational efficiency, reliability, process competitiveness and diversifying energy sources. The most important points to highlight in 2011 are: (i) energy processes; (ii) efficiency; (iii) reliability; and (iv) energy diversification.

Related Information

Environmental Impact Assessment: This portal contains the documents on the Environmental Impact Assessments conducted for Ecopetrol's projects feasibility. Studies prior to this date must be requested through the Citizen Participation Office or can be viewed at the Ministry of Environment and Sustainable Development.

Here you can see the fields, refineries, ports, pumping and storage stations, as well as the transportation network that conforms the country's oil infrastructure. It includes property from Ecopetrol, from third parties that have Current Association Contracts and from other private companies with operation in Colombia.

LUKOIL Kaliningradmorneft, realising the responsibility for environmental safety and rational use of natural resources, organised in 2004 a service for operational satellite monitoring of oil spill pollution in the southeastern Baltic Sea. In total, 86 large oil spills were detected in 88 ASAR ENVISAT images received in June and December 2004. Thirty kilometers area around the D6 oil platform is free of oil spills. Main sources of oil pollution are ships. For auxiliary needs, 617 IR and VIS AVHRR (NOAA) and MODIS (Terra and Aqua) images were processed and analysed, as well as 110 maps of near surface wind derived from the QuikSCAT scatterometer and 73 maps of wave height derived from the JASON1 altimeter were constructed. About 200 oil spills (real and virtual) drift forecasts were done basing on the numerical model Seatrack Web (SMHI). The results obtained in 2004 show the efficiency of complex satellite monitoring of ecological state in the southeastern Baltic Sea, which will be continued in the following years.

Case of Oil India Limited (OIL)

Oil India Limited (OIL) is an Indian company. The company's products and services include crude oil, natural gas, LPG and pipeline transportation. The company holds 1,560,890 square kilometres of acreage, including those in India and overseas, covering 78 blocks, of which it holds in India 13 New Exploration Licensing Policy (NELP) as operator, 1 NELP as joint operator, 19 NELP as non-operator, 2 as joint venture (JV), 8 nominated petroleum exploration licence (PELs), 1 CBM block and 21 petroleum mining lease (PMLs). The company holds 3 blocks as operator, 8 as non-operator and 2 as JV partner overseas. The company's 50 per cent owned JV Indoil Netherlands B.V. is holding in exploration, production, marketing, trade, transport and extraction of oil, gas, hydrocarbons and minerals.

Core Purpose

The fastest growing energy company with a global presence providing value to the shareholder.

Vision

- OIL is the fastest growing energy company with highest profitability.
- OIL delights the customers with quality products and services at competitive prices.
- OIL is a learning organisation, nurturing initiatives, innovations and aspirations with best practices.
- OIL is a team, committed to honesty, integrity, transparency and mutual trust creating employee pride.
- OIL is fully committed to safety, health and environment.
- OIL is a responsible corporate citizen deeply committed to socioeconomic development in its areas of operations.

In 1981, OIL became a wholly owned Government of India (GoI) enterprise. Today, OIL is a premier Indian National Oil Company engaged in the business of exploration, development and production of crude oil and natural gas, transportation of crude oil and production of LPG. OIL also provides various E&P related services and holds 26 per cent equity in Numaligarh Refinery Limited. The authorised share capital of the company is ₹500 crore. The issued, subscribed and paid share capital of the company is ₹240.45 crore. At present, The GoI, the Promoter of the Company is holding 78.43 per cent of the total issued & paid-up capital of the company. The balance 21.57 per cent of the OIL equity has over 1 lakh square kilometres of PEL/ML areas for its exploration and production activities, most of it in the North-East Indian, which accounts for its entire crude oil production and majority of gas production. Rajasthan is the other producing area of OIL, contributing 10 per cent of its total gas production. Additionally, OIL's exploration activities are spread over onshore areas of Ganga Valley and Mahanadi. OIL also has participating interest in NELP exploration blocks in Mahanadi offshore, Mumbai deepwater, Krishna-Godavari deepwater, etc., as well as various overseas projects in Libya, Gabon, Iran, Nigeria and Sudan. In a recent CRISIL-India Today survey, OIL was adjudged as one of the five best major public sector undertakings (PSUs) and one of three best energy sector PSUs in the country.

Corporate Governance

Detailed below is the Code of Conduct for board members and senior management personnel of OIL:

- Introduction to the code of conduct: This Code of Conduct (hereinafter referred to as 'Code') is applicable to the Members of the Board (hereinafter referred to as 'directors') and the Senior Management of OIL (the 'company'). It reflects the company's underlying ethical values and commitment to lay standards of integrity, transparency, fairness, accountability and pursuit for excellence.

(i) The purpose of this Code is to demonstrate the ethical and transparent process in managing the affairs of the company, and thus to sustain the trust and confidence reposed in the management by the stakeholders and business partners. Directors and senior management are expected to understand, adhere to, comply with and uphold the provisions of this Code and the standards laid down hereunder in their day-to-day functioning.

(ii) The principles prescribed in this Code are general in nature and lay down broad standards of compliance to Clause 49 of the Listing Agreement with the stock exchanges, where the shares of the company are listed for trading.

(iii) The company currently has in place the OIL Executive Conduct, Discipline & Appeal Rules (CDA Rules)/Standing Orders as amended from time to time which govern the conduct of all employees of the company including whole-time directors. The rules/orders provide for good conduct and discipline as well as provide for penalties in the case of violations and the appeal mechanism.

(iv) It is clarified that this Code is applicable to all directors and senior management, including the part-time Directors. In case of whole-time directors and the senior management personnel, this Code is applicable over and above the CDA rules of the company.

(v) The Code comes into force with effect from the date of listing of shares of the company on the stock exchange.

(vi) All directors and senior management shall sign the acknowledgment form and return the form to the Company Secretary indicating that they have received, read and understood the provisions of the Code and agree to comply with the same. All directors and senior management shall be required to affirm compliance with this Code on an annual basis, within 30 days of close of every financial year to the company secretary.

- Definition and Interpretation: In this Code, unless repugnant to the meaning or context thereof, the following expressions, wherever used in this Code, shall have the meaning assigned to them below:

  'Directors' shall mean all the members of the Board of Directors of the Company, including the part-time Directors. 'Company' shall mean OIL. 'Whole-time Directors' shall mean the board members who are in whole-time employment of the company. 'Part-time Directors' shall mean the board members who are part-time directors and not in the whole-time employment of the company. 'Senior management' shall mean executives one level below the board level. In this Code, words importing masculine shall include feminine and words importing singular shall include plural or *vice versa*.
- Applicability: This Code shall be applicable to the following persons: (i) all directors and (ii) senior management.
- Key Requirements: The directors and senior management shall act within the authority conferred upon them in the best interests of the company and will:
  - (i) act in the best interest of and fulfil their fiduciary obligations to the company and its shareholders;
  - (ii) act honestly, fairly, ethically and with integrity;
  - (iii) conduct themselves in a professional, courteous and respectful manner and not take improper advantage of their position;
  - (iv) act in good faith, responsibly, with due care, competence and diligence, without allowing their independent judgement to be subordinated;
  - (v) use their prudent judgement to avoid all situations, decisions or relationships which give or could give rise to conflict of interest or appear to conflict with their responsibilities within the company;
  - (vi) not exploit for his/her own personal gain, opportunities that are discovered through use of corporate property,

information or position, unless the company declines to pursue such opportunity for its business interest(s);

(vii) avoid conducting business on behalf of the company except with the prior approval of the board with: (a) a relative; (b) a private limited company in which he or his relative is a member or a director; (c) a public limited company in which he or his relative holds 2 per cent or more shares or voting right; and (d) with a firm in which the relative is a partner;

(viii) disclose and avoid having any personal and/or financial interest in any business dealings concerning the Company;

(ix) avoid any dealings with a contractor or supplier that compromises the ability to transact business on a professional, impartial and competitive basis or influence decision to be made by the company;

(x) not hold any positions or jobs or engage in outside businesses or other interests that are prejudicial to the interests of the company; and

(xi) inform the board, at the earliest opportunity, any existing or potential conflict of interest situation.

It is clarified that it would not be a conflict of interest for the director and senior management of the company or members of their immediate family to have dealings in normal course with and to obtain services from persons or entities who also deal/provide services to the company, including legal, accounting or brokerage services, loans and services from banks or insurance from insurance companies, etc., at rates customary for similarly situated customers.

- Other Directorships: Unless specifically permitted by the board, the board members shall not serve as director of any other company or as partner of a firm that is engaged in a competing business with the company. This clause is not applicable to ex-officio government nominee directors. Senior management personnel shall obtain prior approval of the chairman and managing director/competent authority for accepting directorship of any other company or partnership of a firm.

- Gifts and Donations: The directors and the senior management shall neither receive nor offer or make, directly or indirectly, any gifts, donations or comparable benefits which are intended to or perceived to obtain undue favours in the transactions detrimental to the interest of the company, except for nominal value, which is customarily given and is of commemorative nature for special events and should never be of a kind that could create an appearance of impropriety.
- Public Representation: The company honours the information requirements of the public and its stakeholders. In all, its public appearance with respect to disclosing information in relation to the company's activities to public constituencies such as the media, the financial community, employees and shareholders, the company or the Group shall be represented only by specifically authorised directors and employees. It will be the sole responsibility of these authorised representatives to disclose company-related information.
- Insider Trading: The directors and senior management shall comply with the Code of Internal Procedures and Conduct in dealing with the securities of the company as applicable from time to time.
- Confidentiality of Information: Any information concerning the company's business, its customers, suppliers, etc., to which the directors and senior management have access or which is possessed by the directors and senior management, must be considered privileged and confidential and should be held in confidence at all times and should not be disclosed to any person, unless: (i) specifically authorised or (ii) the same is part of the public domain at the time of disclosure or (iii) is required to be disclosed in accordance with applicable laws.

- Regulatory Compliance: Every director and senior management shall, in his or her business conduct, comply with all applicable laws and regulations, both in letter and in spirit, in all the territories in which he or she operates. If the ethical and professional standards set out in the applicable laws and regulations are below that of the Code, then the standards of the Code shall prevail.
- Health, Safety and Environment: The Directors and Senior Management shall be committed to prevent the wasteful use of natural resources and minimise any hazardous impact of the development, production, use and disposal of any of its products and services on the ecological environment.
- Protection of Assets: The Directors and Senior Management of the Company shall employ the assets of the Company for the purpose of conducting the business for which they are duly authorised. These include tangible assets such as equipment and machinery, systems, facilities, materials, resources as well as intangible assets such as proprietary information, relationships with customers and suppliers, etc.
- Amendments to the Code: The provisions of this Code can be amended/modified by the Board of Directors of the company from time to time in line with any changes in law, changes in company's philosophy, vision, business plans or otherwise, and all such amendments/modifications shall take effect from the date stated therein. All directors and senior management shall be duly informed of such amendments and modifications.
- Placement of the Code on Website: Pursuant to Clause 49 of the Listing Agreement, this Code and any amendments thereto shall be posted on the website of the company.
- Enforcement of Code of Conduct: All directors and/or senior management shall be accountable for fully complying with this Code.
- Consequences of Non-Compliance of this Code: In case of breach of this Code by the directors, the same shall be dealt

with by the Board of Directors for initiating appropriate action, as deemed necessary. In case of breach of this Code by the senior management, the same shall be dealt with by Chairman and Managing Director in accordance with the CDA rules of the company.

Corporate Social Responsibility: The OIL Way

Since its inception, OIL has always given top priority towards the all-round development of the people residing in and around the Company's area of operation. OIL management understands that there is a need to strike a balance between the overall objectives of achieving corporate excellence *vis-à-vis* the corporate responsibility towards the community. It is this twin objective of business and social commitment that has prompted OIL to embark upon massive programmes of educational, health and infrastructural development endeavours.

Vision Statement

"OIL is a Responsible Corporate Citizen deeply committed to socio-economic development in its areas of operations."

Philosophy

"To promote the awareness and practice of Good Corporate Citizenship, by business becoming an integral part of societal process where people have access to resources to make informed choices and decisions towards a more humane and compassionate society."

Strategy

"To mobilise core competencies and resources of business, public organisations and Government Institutions...to facilitate their working in partnership on projects that benefit communities."

Resource Allocation: From a relatively small budget of ₹ 20 lakh in the early 1980s, OIL currently spends a minimum of 0.75 per cent of its net profit on CSR.

A Pat on the Back: OIL's exemplary achievements in CSR have earned the Company due recognition from prestigious organisations.

The World Council for Corporate Governance (WCFCG) and Centre for Corporate Governance (CFCG) awarded OIL with the prestigious Golden Peacock Award for Corporate Social Responsibility in 2002. OIL also bagged the prestigious TERI Corporate Social Responsibility Award, 2004, which was presented to OIL in recognition of corporate leadership for good corporate citizenship and sustainable community development initiatives. The search for hydrocarbons is on and technology has always been at the core of OIL's activities. Induction of Information and Communication Technology as an enabler to the core functions of OIL has been a driving force. OIL's IT initiatives through various software solutions, state-of-the-art servers, optical fibre, satellite up-linking, terminals, platforms programme developers, digital data, networking, workflow have been enabling OIL to sail along the information superhighway, empowering employees and management to take techno-economic decisions so vital in today's era of cut-throat competitions. OIL's IT Department is responsible for providing a proper workflow where people have access to latest technology, which ensures accurate, timely and consistent transmission of data.

Summary of Various CSR Initiatives

Social Survey: OIL conducted a social survey through Dibrugarh University in the year 1983 (in its operational areas in the North East) to evolve a CSR Strategy. Based on the recommendation of the survey, the company introduced a scheme entitled Social Welfare Programme (SWP) in 1984 and later on another scheme called Areas Development Scheme (ADS) in 1996. The SWP policy places OIL's commitment of protecting the environment, upliftment of education, health and socioeconomic development the topmost priority. All-round development of education, rural infrastructure, primary health care, environment protection, promotion of sports/sports person (especially rural sports), assistance to youth and women organisations in carrying out community development activities are some of the key areas covered under SWP. ADS was introduced in 1996 with the objective of supporting the various socio-cultural activities in the operational areas. The scheme covers the construction of roads, setting up of educational institutions and primary health centres in the North East region and other operational areas of the company. Based on the recommendations of the social survey, OIL has till date constructed over 100 bridges/1,400 kilometres of roads public Works Department (PWD/village roads) and provided financial assistance to most of the educational institutions in and around OIL's operational areas in Assam.

Health: OIL's Mobile Dispensary Services is a mobile hospital. The mobile hospital services of OIL have always been one of the most significant community welfare projects of the Company. Today, the team of doctors and paramedical staff of OIL hospital cater to the basic medical needs of the villages in and around OIL's operational areas.

In the recently carried out social audit of OIL's community welfare initiatives, the Company's mobile hospital services were ranked very high by all the respondents. Based on this feedback, the Company decided to increase the number of visit as well as extend the services to new areas. From the earlier 16-20 visits per month, the mobile services make 40-50 visits per month. The Company has signed a memorandum of understanding (MoU) with St Luke's Hospital, which will extend the services to villages in Tinsukia District while the company's own team would cater to villages in Dibrugarh District.

Education: Since inception, OIL has been providing financial assistance to all-round development of education from primary to higher education. As of the day this chapter was written, the company has provided financial assistance towards construction of school building, libraries, auditoriums, teachers' common room, etc., in over 350 institutions in the company's operational areas in upper Assam. The company also provided merit scholarships to students who excel in their 10th and 12th board exams from schools in and around OIL's operational areas in upper Assam. In its commitment to the all-round development of higher education, OIL has sponsored two prestigious Chairs (K.D. Malaviya Chair in Geosciences at Dibrugarh University and Sankardeva Chair in Literature at Guwahati University). Recently, the Company has extended a one time financial assistance of ₹ 3.68 crore to the Business and Centre of Management Studies, Dibrugarh University, and ₹ 1.5 crore to Assam Engineering College, Guwahati.

General Nursing Midwifery (GNM) Training in OIL Hospital, Duliajan: The nursing school in OIL Hospital, Duliajan, conducts a three year General Nursing Midwifery (GNM) training course which is recognised by the Directorate of Medical Education, Government of Assam. The annual intake is 20 candidates, out of which two seats candidates each are reserved for scheduled caste (SC) and scheduled tribe (ST) and four for other backward class (OBC) candidates. Stipend is paid to the students in addition to limited hostel accommodation, uniform and protective clothing.

Financial Assistance to Socio-cultural/Sports/Youth Organisations / Mahila Samities: The company has supported the youth and women organisations in a big way. The support is mainly in terms of financial assistance towards enabling the different organisations in carrying out their developmental activities.

OIL places a heavy accent on sports. Besides sponsoring national ranking tournaments every year, the company recruits promising young players and gives them every opportunity to advance their sports careers. OIL has taken up schemes of holding exhaustive football camps with trained coaches to provide professional training in the rural areas of upper Assam. OIL is also promoting rural sports by developing several village playgrounds and organising different games in the oil field areas.

Initiatives for Socioeconomic Development

Project Rupantar: The New CSR Initiative: In order to address the problem of growing unemployment and poverty, OIL has undertaken a long-term project named Rupantar (an Assamese word meaning transformation) of investing in projects, which can help the unemployed youths to find alternate source of employment. OIL signed an MoU with the State Institute of Rural Development (SIRD), Assam, at the company's field headquarters in Duliajan on 8 September 2003. The guiding provisions of the MoU are a long-term vision to generate sustainable self-employment avenues for the educated unemployed youths in and around the company's operational areas in upper Assam. The project aims to create self-employment avenues and promote entrepreneurship in the region and generate sustainable sources of livelihood. In order to help the large number of unemployed youth and strengthen the rural economy, the focus is on development of agro-based industries, diversification in handloom products with special focus on *eri* and *muga* (world famous golden silk of Assam), poultry farming, pig breeding, duck rearing, fishery, sericulture, organic farming, etc. Till date over 2000 self-help groups (SHGs) have already been formed and a large number of these SHGs have received loans from banks.

The agro-based industries are not only helping in ensuring the economic independence of the people residing in and around OIL's operational areas but at the same time result in wealth creation and self-employment to a large section of the people of upper Assam. Oil India christened this new project as Rupantar, which is providing the rural poor and the unemployed an opportunity to become self-employed.

Oil India Rural Development Society (OIRDS)—Agriculture Development Programme: Oil India Rural Development Society (OIRDS) under the aegis of OIL adopts villages under registered Pathar Parichalona Samity (PPS), and provides assistance and technical guidance to farmers in nearby villages of OIL's operational areas for agricultural development. Extensive in-field training on modern methods of cultivation both in *kharif* and *rabi*, farming equipment, high-yielding seeds, organic fertilisers, etc., are introduced and distributed to farmers of the region. Introduced in the year 1991, 31 villages have been adopted under this scheme till date.

Handicraft Training and Production Centre (HTPC): Established in 1984, OIL's Handicraft Training and Production Centre, located at Duliajan imparts training to women on handicraft, weaving, embroidery and tailoring on monthly stipendiary for a period of nine months period. Till date, over 500 women have been trained, most of whom have set up their own small production units.

Project ASWANOIL: The project is replacing the telecommunication system of 1,157 km crude oil trunk pipeline which supplies crude oil to the four refineries in India's north eastern region with OFC. The additional capacity (bandwidth) that will be available in the OFC system would act as an infrastructure provider for value-added services in telecommunication business. OIL has committed to provide the spare bandwidth to the Government of Assam's Assam State Wide Area Network (ASWAN) project which aims to connect all the district headquarters of the state.

Case of Oil and Natural Gas Corporation Limited (ONGC)

Oil and Natural Gas Corporation Limited (ONGC) is an Indian company. The Company is mainly engaged in the oil exploration and production activities. The Company operates in two segments: Exploration, and Production and Refining. As of 31 March 2012, the Company had made 23 discoveries in domestic fields: 15 new prospects and 8 new pool discoveries. Out of the 15 new prospect discoveries, 7 are in NELP blocks. It is operating in four coalbed methane (CBM) projects, which include Jharia, Bokaro, North Karanpura and Raniganj. During the fiscal year ended on 31 March

2012, the company had a crude oil production of 38.09 million metric tonnes and natural gas production of 47.56 million metric tonnes. Its subsidiaries include ONGC Videsh Limited (OVL), Mangalore Refinery & Petrochemicals Ltd., ONGC Nile Ganga BV (ONGBV), ONGC Nile Ganga (Cyprus) Ltd., Jarpeno Limited, Imperial Energy Corporation Plc, Imperial Energy Limited and Imperial Energy Kostanai Limited.

Sustainable Development Policy

ONGC believes that sustainable development requires contribution from all societal players and significantly so from corporate. The sustainable development policy is aimed at driving efficient and effective implementation of sustainable development activities, initiatives and projects across ONGC. It considers the sustainable management of water, materials and energy, addressing climate change through carbon management as the key broad responsibilities towards environmental sustainability. Its vision is to gradually work towards reducing its carbon and water footprint, innovative beyond compliance management of waste and prudent energy management and biodiversity conservation.

Its approach for working on our sustainable development responsibilities will be through projects in the specified areas and will be driven by an adequately empowered organisational structure in ONGC with a system of management oversight, review and control.

Environment Sustainability

According to ONGC, corporate sustainability report, 2013, an ideal route to cover the elements of its business-specific sustainable development issues across the environmental dimension is to remain focused on carbon management. It considers mitigation of global GHG from operations as a critical area of environmental sustainability.

Sustainability Challenges and Issues

With the intention of ensuring improved sustainability management, monitoring and reporting process, and progressively integrating sustainable development in the organisational working, it has identified issues that are most important for its stakeholders. Towards this, it followed the GRI framework and followed structured engagement process.

The engagement process (formal and informal) involved identification of key stakeholders (employees of ONGC, GoI, business partners and community) and subsequently understanding major sustainability issues. Considering the diverse nature of operations and concomitant diverse aspirational levels of the employees engaged in these operations, it has been proposed that systematic engagement with its employees should be the focus for the purpose of the current reporting. Based on its materiality determination exercise for sustainability, a macro level mapping of issues for the organisation in the overall business and sustainability context with relative priority in terms of management attention, plans and work programme during the reporting period is presented below. Elements of this issue map, which have found focused management attention and have moved up in the priority hierarchy during the reporting period, are energy security and reserve accretion; worker health, safety and asset security; water and waste; hiring and retaining workforce; and community. The issues of innovation and R&D, and ethics have also emerged as sustainability issues this year and would be addressed after due diligence.

National Gas Star Programmme

Towards identifying, quantifying and reducing the fugitive methane emissions from our production operations, ONGC entered into an MoU with the USEPA in August 2007, to undertake Methane to Market (now GMI) projects in ONGC under its Natural Gas STAR

Programme to promote development, implementation and reporting of profitable, voluntary methane emission mitigation activities.

The technical collaboration between the USEPA and the ONGC under the MoU, administered under the GMI partnership, has built a strong base of knowledge and capacity within ONGC to cost-effectively reduce methane emissions now and into the future.

As of now, ONGC has reduced approximately 10.54 MMSCM, (million metric standard cubic metres) since joining the partnership in 2007 which is equivalent to reducing 150,000 from the atmosphere and has chalked out an elaborate plan to map all our production installations for fugitive hydrocarbon emission and make the installations leak free in the coming few years. As of now, it has completed leak detection and measurement at 56 relevant facilities. Once achieved its planned objective, it would be a remarkable feat for ONGC and the oil & gas industry at large in India.

Sustainable Water Management

ONGC is working on a company-wide 'Sustainable Water Management Strategy' aimed at reduction in specific fresh water consumption and reporting on the water footprint based on internationally recognised standards and practices. Its action plans include baseline assessment of water use, build reporting capability in the short term followed by operation-specific sustainable water management plans, location specific standard operating procedures (SOPs) with water recycling, and reuse targets as appropriate in the medium to long term.

Energy Efficiency

ONGC has a mission to create an energy centre focused on conducting research on marketable solutions in alternate energy. The genesis of the centre is from the growing realisation that oil production is likely to peak soon and there is a need to find implementable alternate energy solutions. This realisation has significant implications for ONGC's business and for the nation at large.

ONGC has set up the ONGC Energy Centre (OEC) for research in renewable and alternate energy sources beyond hydrocarbons, which can be brought to market. The objective is in line with the vision of ONGC to become an integrated energy company and not just restrict itself to oil and gas. The centre was pioneered by former Chairman and Managing Director Late Mr Subir Raha with motto "Mind to Market." The ONGC Energy Centre aims to conduct research in alternate energy specifically focusing on projects which have the potential to make an impact on India's energy scene. The centre does not stop at just R&D, but pursues beyond to 'demonstrate for commercialisation.'

The OEC has a unique operating model—minimal full-time employees, project-based teams, flexible compensation structure, and a culture that supports research, creativity, renewal and transmigration of ideas.

The OEC is headed by a Director-General and has three organisational units: Technical Research Unit, Research Advisory Unit and Administrative Support Unit. The Technical Research Unit is the key operational unit responsible for implementation and coordination of research projects. The Research Advisory Unit is responsible for generating new ideas, evaluation and selection of research projects by providing high quality decision-critical information. There is an Advisory Committee comprising of eminent personalities with expertise and experience in various fields. The committee screens the various R&D proposals and also monitors the progress of the projects. Presently, the OEC has its offices in Delhi, Ahmedabad and Mumbai (Panvel).

OEC has been implementing research projects in several new and alternative sources of energy for the last several years. These projects are in advanced stages of implementation, in collaboration with various national and international academic, research and industrial organisations. The salient features of the on-going research projects are as follows.

Hydrogen Generation Through Thermochemical Processes

Hydrogen is a clean energy carrier. It is abundant on earth, but in combined form in several compounds, the most common being water. However, producing hydrogen from water is energy intensive. It is imperative to research methods to generate hydrogen from water in an energy-efficient way.

OEC has taken initiative to work on development of thermochemical processes for decomposition of water utilising waste nuclear and/or solar heat. From among various possible options, development of closed-loop copper-chlorine (Cu-Cl) and closed/ partially open-loop iodine-sulphur (I-S) cycles have been taken up under this project.

Exploration for Uranium

ONGC has so far drilled more than 10,000 wells, many of which have shown the presence of uranium on gamma ray spectrometer (GRS) logs. Secondary uranium mineralisation at shallow depths is reflected on spectral gamma ray log or in its absence, by high gamma ray counts on gamma ray log, which is one of the essential logs of the logging suite, recorded in wells drilled for exploration/exploitation of hydrocarbons. Secondary uranium mineralisation at shallow depths has been observed in several petroliferous basins in the world.

OEC has undertaken a project to analyse the well log data of ONGC to delineate locations of probable uranium mineralisation in the sedimentary basins of India. Parametric wells are drilled at identified locations, cored and logged for spectral gamma ray for confirmation of the presence of uranium, amenable to extraction from host rock by *in situ* leaching (ISL) technique. The cores, recovered from zones of interest, are analysed in laboratories through geological, geochemical and radiological methods, for determining the actual uranium concentration and its leachability.

OEC has entered into an MoU with the Atomic Minerals Directorate, Hyderabad, for these activities.

Solar Thermal Project

Solar energy can be converted to heat and/or electricity using solar thermal or photovoltaic route. There are several methods to generate electricity using solar thermal route. One of the methods is using a combination of solar dish with a stirling engine to produce electricity. Stirling engine is considered to be one of the most efficient heat engines. One of the advantages of using these engines is that no water is required.

OEC is currently working on field performance evaluation of solar thermal dish stirling engine system, developed by a US company. Three units, which generate grid quality alternating current AC electricity of 230 V, 50 Hz, are installed and operational at the Solar Energy Centre of Ministry of New and Renewable Energy at Gurgaon.

Bioconversion of Lignite to Methane

With the increasing demand for energy for development and the growing concern for climate change and the need to limit GHG emissions, it is necessary to find out efficient and cleaner ways to utilise our coal reserves. There are also many coal seams, which cannot be mined. There are several types of microbes, acting in tandem (as a consortia), which can generate methane from coal and oil. A unique consortium of microbes is required for each coal seam/ oil reservoir. Research is underway to identify the specific consortia that can act on different seams/reservoirs.

OEC had taken the initiative to undertake research on development of biotechnological process for conversion of lignite/ coal to methane gas and humic acid. Humic acid, which increases the availability of soil nutrients, is a good plant growth enhancer and thus finds wide application in agriculture and horticulture.

A collaborative project with Indian Institute of Technology (IIT), Kharagpur to generate methane and humic acid from lignite in surface conditions as well as *in situ* is in progress. The biotechnology for production of methane and humic acid from lignite has been developed in the laboratory. Based on the developed biotechnology, mini-pilot scale studies for surface production of methane and humic acid are in progress.

Kinetic Hydro Power

There are extensive opportunities to generate electric power using kinetic energy available in any flowing water stream, without construction of dams for storage of water. Floating turbines can be used to harness the kinetic energy of flowing streams of rivers or tail-race of existing dams to generate electricity for powering un-electrified homes as well as meeting distributed power requirement for agriculture and industrial applications.

OEC is working with M/s Natural Power Concepts (NPC), US, for field trials of their recently developed In-stream Auger Turbine (IAT) in Indian conditions. On successful completion of field trials, these turbines will be deployed in India.

Geothermal Energy

Geothermal energy can be harnessed by extracting heat from the earth by drilling water or steam wells in a process similar to drilling for oil. During the past decade, the feasibility of generating electricity from relatively low-temperature geothermal resources in sedimentary basins has been demonstrated in many countries. Deep and hot aquifers in the sedimentary basin can be used as source of geothermal energy.

In western India, the Cambay basin is underlain by the Deccan trap and has geothermal gradient, with many known permeable formations at different depth levels, which can be utilised for harnessing geothermal energy. A Belgium-based company M/s.

Talboom is collaborating with OEC to set up a pilot power project in Cambay basin. Extensive data available from the Cambay basin have been studied to identify the target area for the pilot. The subsurface heat flow modelling in the identified area is under way. The location for the first well will be finalised soon and the drilling of the well is planned this year.

Therefore, it is evident from the cases presented above that the corporate world conscious of social and environmental sustainability practices, and has policies and procedures in place towards their implementation. The approaches of the corporates clearly demonstrate sensitivity in handling social and environmental issues. All the five cases demonstrated an urge towards energy efficiency through diversification and also undertaking research and development activities.

CASE POINT

Montara Oil Spill in the Timor Sea: "A Failure of Sensible Oil Field Practice"

Exploration and production activities in the oil and gas industry continue to push the extremes—deeper water, harsher environments, political and regulatory uncertainty, and skilled labour shortages. The Montara oil spill in the Timor Sea and the Macondo oil spill in the Gulf of Mexico both contain lessons that can reduce operational risk. Regulators are moving in the direction of requiring real time data from the platform to shore to proactively manage these risks.

While British Petroleum (BP's) deepwater spill in the Gulf of Mexico was worldwide news for months, the worst incident of its kind occurred in Australia in August 2009 where oil leaked for 74 days before the well could be brought under control. Yet, in Australia media interest was muted after the first few days and most people even here would have forgotten that it ever happened. Nobody seems to care if you don't have a household name like BP?[9]

The Montara Development Project is owned and operated by PTTEPAA, a subsidiary of the Thai company PTT Exploration and Production Public Company Limited (PTTEP). The Development is located in a remote area of the Timor Sea, approximately 250 km north-west of the Western Australian coast, and almost 700 km from Darwin.

The Montara oil spill occurred after a blowout and fire on the Montara WHP (wellhead platform). The blowout occurred on the H1 well on 21 August 2009 while the West Atlas jackup drilling rig was operating over another well at the time. Sixty-nine workers were safely evacuated from the drilling rig. The leak continued for 74 days, until an intervention well was successful on its fifth attempt on 3 November and mud was pumped in to 'kill' the well.

The Thai operator estimated that the flow of oil may have been between 1000-1500 barrels per day in the early stages and declined to 400 barrels per day later. In total, it is estimated that 30,000 barrels were spilled (plus presumably a good deal of gas and/or condensate). The total surface area

9. Montara Oil Spill: "A failure of sensible oilfield practice" Posted by Phil Hart on December 8, 2010 - 11:25am. Topic: Geology/Exploration Tags: Australia, Montara, oil spill [list all tags]

over which oil or sheen was observed at one time or another was around 90,000 square kilometres.

From the inquiry: at the time the H1 well was suspended in March 2009, not one well control barrier complied with PTTEPAA's own Well Construction Standards (or, importantly, with sensible oilfield practice).

Relevantly, the 9⅝" cemented casing shoe had not been pressure tested in accordance with the company's Well Construction Standards, despite major problems having been experienced with the cementing job. In particular, the cement in the casing shoe was likely to have been compromised as it had been substantially over-displaced by fluid, resulting in what is known as a 'wet shoe.' None of this was understood by senior PTTEPAA personnel at the time, even though the company's contemporaneous records, such as the *Daily Drilling Report* (DDR), clearly indicated what had happened. The multiple problems in undertaking the cement job—such as the failure of the top and bottom plugs to create a seal after 'bumping,' the failure of the float valves and an unexpected rush of fluid—should have raised alarm bells. Those problems necessitated a careful evaluation of what happened, the instigation of pressure testing and, most likely, remedial action. No such careful evaluation was undertaken. The problems were not complicated or unsolvable, and the potential remedies were well known and not costly. This was a failure of 'sensible oilfield practice 101.'

As on April 2009 when the H1 well had been suspended and the West Atlas rig had departed from the Montara WHP to undertake other work, not one well control barrier in the H1 well had been satisfactorily tested and verified, and one barrier that should have been installed was missing. In other words, the H1 well was suspended without regard to PTTEPAA's own Well Construction Standards or sensible oilfield practice.

When the West Atlas rig returned to the WHP in August 2009, it was discovered that the 13⅜" PCCC (pressure containing anti-corrosion caps) had never been installed. The absence of this PCCC had resulted in corrosion of the threads of the 13⅜" casing and this, in turn, led to the removal of the 9⅝" PCCC in order to clean the threads. This was viewed by PTTEPAA personnel as a mere change of sequence that simply involved bringing forward the time of the removal of the 9⅝" PCCC. PTTEPAA's Well

Construction Manager, Mr Duncan, took a positive decision not to reinstall the 9⅝" PCCC. This meant that, according to PTTEPAA's operational forecast and drilling programme, the H1 well would have been exposed to the air without any secondary well control barrier in place for some 4-5 days, with sole reliance on an untested primary barrier (the cemented 9⅝" casing shoe) that had been the subject of significant problems during its installation.

After the 9⅝" PCCC had been removed, the H1 well was left in an unprotected state (and relying on an untested primary barrier) while the rig proceeded to complete other planned activities as part of batch drilling operations at the Montara WHP. The blowout in the H1 well occurred 15 hours later.

It seems that Halliburton who performed the cement job was aware that there was a problem, but the operator accepted it anyway. Despite the fact that the Halliburton cementing report clearly showed that the casing shoe could not be regarded as having barrier integrity, the report was 'signed off' by Mr Treasure (PTTEPAA's senior on-rig representative) later in the evening on 7 March 2009, with a handwritten annotation "good job well done."

In this case, it seems that the lines of responsibility for accepting the poor cement job are clear and that Halliburton does not bear much of the blame. Whether that will the case for the Gulf of Mexico incident remains to be seen. One doubt that Halliburton has pleased their name has been associated with both incidents, but they do not have many peers in their line of work so it could easily be coincidence.

In summary, as at April 2009 when the H1 well had been suspended and the West Atlas rig had departed from the Montara WHP to undertake other work, not one well control barrier in the H1 well had been satisfactorily tested and verified, and one barrier that should have been installed was missing. In other words, the H1 well was suspended without regard to PTTEPAA's own Well Construction Standards or sensible oilfield practice.

In essence, the way that PTTEPAA operated the Montara oilfield did not come within a 'bulls roar' of sensible oilfield practice. The blowout was not a reflection of one unfortunate incident, or of bad luck. What happened with the H1 well was an accident waiting to happen; the company's systems and processes were so deficient and its key personnel so lacking in basic competence, that the blowout can properly be said to have been an event waiting to occur. Indeed, during the course of its public hearing, the Inquiry discovered that not one of the five Montara wells currently complies with the company's Well Construction Standards. Indeed, so poor has PTTEPAA's performance been on the Montara oilfield, that the inquiry considered it imperative that remedial action be instituted.

Offshore Industry Regulation

The Northern Territory is too small to manage the regulation and oversight of the oil industry, and a new national regulator has been recommended—this would seem well overdue.

The Northern Territory has also contended that "at all material times prior to the [Blowout], the Territory appropriately administered the licence area within which the Montara Wellhead Platform is located." The inquiry has no hesitation in rejecting this contention.

And again..

The inquiry is of the view that nothing should detract from the primary responsibility of PTTEPAA to ensure well integrity. However, the inquiry finds that the NT DoR's regulatory regime was totally inadequate, being little more than a 'tick and flick' exercise.

By way of example, when PTTEPAA submitted an application to suspend the H1 well utilising PCCCs rather than a cement plug, it received preliminary approval in 30 minutes. However, as the inquiry heard from the manufacturers of the particular PCCC used, they were not intended to be used as barriers against a blowout. In this respect, the information that had been conveyed to the Department of Resources, Northern Territory was seriously deficient. However the NT DoR, which had no real prior experience with PCCCs, gave almost immediate approval for their use.

Government Response: The 21 August 2009 uncontrolled oil and gas release at the Montara oilfield, operated by PTTEP Australasia (Ashmore-Cartier) Pty Ltd (PTTEP AA), and the more recent incident on 20 April 2010 at the BP operated Macondo oilfield in the Gulf of Mexico, where 11 lives were lost, serve as strong reminders to governments, regulators, the offshore petroleum industry and the broader community of the risks of complacency in the operation and regulation of offshore petroleum activities.

As the report states, regulation of the Australian oil industry needs a complete overhaul to prevent wildcat operators like this creating more 'accidents waiting to happen.' I am personally amazed that an oil company as bad as this still exists and that they were allowed to operate here in Australia, a country that is normally seen as modern and effectively regulated.

In this context, the most significant finding of the inquiry, which has been agreed to by the Australian Government, is the creation of a new national regulator to oversee the offshore petroleum industry. Hopefully they will have the resources and skills to prevent such an incident, which clearly the Northern Territory agency did not have.

Conclusion

While the Montara spill was perhaps smaller in order of magnitude smaller than the Macondo Gulf of Mexico spill, it is surprising how little attention was paid to Montara even in Australia. It could be argued that the media overplayed the BP incident (serious as it was) with seemingly 24 hours non-stop coverage and hyperbole. However, for the Australian media and public, to turn an almost blind eye is a little concerning. Perhaps our 'dig it up, ship it out' mentality has weakened our concern for the environment. Or perhaps the lack of a recognisable household name meant there was nowhere for people to turn their anger and attention.

Most of the major international oil companies have had their moment (or extended duration) in the public gaze at one time or another over the last decade. The national oil companies from around the world rarely feel any such heat no matter how low their environmental or ethical standards may fall. Despite evidence of an 'accident waiting to happen,' the criticisms levelled at the Thai operator will not do it much harm outside Australia, especially since they do not exist as far any consumers are concerned. They will though have to work hard not to lose their license to operate in Australia given the numerous and blatant failings uncovered.

However, the review into BP's Gulf of Mexico Macondo blowout will not occur outside the spotlight. The entire industry will be hoping that the mistakes among BP and its partners and contractors do not amount to a 'failure of sensible oilfield practice' or they will all feel the repercussions, through increased costs, delays and even cancelled projects around the globe.

References

Collins, E., J. Roper and S. Lawrence (2010). "Sustainability Practices: Trends in New Zealand Businesses", *Business Strategy and the Environment* 19(8): 479-94.

Collins, Eva, Geeta Duppati and A.K. Rath(2013). "Sustainability Practices, Drivers and Barriers: A New Zealand/India Comparison", *Journal of Commerce & Business Studies* 1.

Mishra, R., S. Sarkar and P. Singh (2012). "Linking CSR to Business Strategy: The Balanced Scorecard Approach", Paper presented at the *Strategic Corporate Social Responsibility: Towards a Sustainable Business*. Hyderabad.

6

Comparing the Similarities and Differences in Corporate Governance Practices and Performance

This chapter also compares the following issues relating to the corporate governance and provides a conclusion.

Corporate Governance

Early comparisons on corporate governance models divided the world into two broad dichotomous systems: the Anglo-American corporate governance system, which is characterised by short-term equity finance, dispersed ownership, strong shareholder rights, active markets for capital control, and flexible labor markets, and the Continental European corporate governance system, which is characterised by long-term debt financing, concentrated blockholder ownership, weak shareholder rights, inactive markets for capital control rigid labour markets (Becht and Roell, 1999; Hall and Soskice, 2001; La Porta *et al.*, 1998; Shleifer and Vishny, 1997). Over time, these stylised dichotomous frameworks have been refined to fit the empirical realities better in different countries.

It is evident from Table 6.1 that with the exception of LUKOIL, the other four oil companies have substantial stake of government equity holdings. With regard to duality, with the exception of the Indian companies, the remaining three companies have the roles of the chairman separated from the chief executive officer (CEO)/

managing director. One commonality is that the board size is above 10 and all the boards have members holding higher and relevant degrees in their fields of expertise.

It is evident from Table 6.1 that the corporate governance model from the context of shareholders investment pattern shows the similarities of the Continental European corporate governance model.

Table 6.1

Corporate Governance: A Cross-country Comparison

Countries	*Board Structures*	*Board Size*	*Shareholding Pattern*	*Qualifications*
LUKOIL	No-duality	Total Size 11 • 9 Independent Directors (IDs) • 2 Executive Directors (EDs)	Investment Managers (11 groups) 99% Government stake is 0%.	All board members are highly qualified in their relevant fields of expertise and hold degrees equal or above masters
Ecopetrol SA	No-Duality	Total 12 • 9 IDs, • 1 Director, Representative of Minority shareholders and • 2 Directors, Representatives of concerned ministries	Holding by Government Agency is 88.49%.	All board members are highly qualified in their relevant fields of expertise and hold degrees equal or above masters
Petrobras Brasilia	No-Duality	Total 10	Holding by federal Govt equity capital is 50.3%	
OIL Ltd	Duality	Total 12 • 1 CMD • 4 EDs • 2 Government Directors/Nominees • 5 IDs	Govt equity capital is 68.43%	All board members are highly qualified in their relevant fields of expertise and hold degrees equal or above masters

contd...

...*contd*...

ONGC Ltd	Duality	Total 12 • 1 CMD • 5 EDs • 2 Govt Directors/ Nominees • 2 Non-official part-time NED Independent	Govt equity capital is 69.23%	All board members are highly qualified in their relevant fields of expertise and hold degrees equal or above Masters

Table 6.2

Financial Performance

ROA	*2011*	*2010*	*2009*	*5-year Average*
LUKOIL	11.64	11.83	10.51	13.06
Ecopetrol SA	20.12	13.59	10.34	17.22
Petrobras Brasilia	6.51	8.18	10.16	10.25
ONGC	10.6	10.25	12.35	12.1
Oil India	14.7	14.95	16.25	16.29
Dividend Yield	*2011*	*2010*	*2009*	*5-year Average*
LUKOIL	4.41	3.39	3.05	3.6
Ecopetrol SA	6.24	3.54	3.66	4.96
Petrobras Brasilia	5.26	2.06	2.57	3.16
ONGC	-3	3	0	1.95
Oil India	na	na	na	na
Tobin's Q	*2011*	*2010*	*2009*	*5-year Average*
LUKOIL	0.45	0.53	0.6	0.63
Ecopetrol	1.95	2.47	1.86	1.94
Petrobras Brasilia	0.21	0.31	0.40	0.38
ONGC	1.11	1.17	0.92	1.19
Oil India	1.486449	1.534626	NA	NA
Efficiency Ratio: Asset Turnover	*2011*	*2010*	*2009*	*5-year Average*
LUKOIL	1.1	1	1	1.06
Ecopetrol SA	0.7	0.6	0.6	0.62
Petrobras Brasilia	0.5	0.4	0.4	0.54
ONGC	0.6	0.5	0.5	0.54
Oil India	0.4	0.4	0.4	0.48

The dividend yield figures indicate dividend policy is important to lure the investors. With the exception of ONGC, the others are doing relatively better. The efficiency ratio for the Indian companies is relatively low when compared to the rest. Likewise, the flagship company of the country is trailing in terms of generating return on assets when compared with others. The Tobin's Q is relatively high for Ecopetrol when compared to the rest of other companies with higher government stake.

Table 6.3

Corporate Social Responsibility and Sustainability

Countries	*Corporate Social Responsibility*	*Sustainability Practices*
LUKOIL	Group social responsibilities consist in an adequate working environment and appropriate remuneration, environmental safety and preservation of the cultural heritage.	– increase petroleum gas utilisation rate through expansion of gas-processing and power-producing capacities; – apply the zero-discharge principle while developing offshore fields; – increase the output of environmentally friendly fuels compliant with the European standards; – comply with greenhouse gas (GHG) reduction provisions of the Kyoto Protocol; – applying power efficient techniques and using alternative energy sources ensure efficient management of natural resources in the areas of LUKOIL Group operations and of those used for production;

contd...

...contd...

Ecopetrol SA	– The strategy of corporate social responsibility (CSR) at Ecopetrol is defined in terms of managing relations with stakeholders in a responsible manner, based on the recognition of their needs and expectations, and the implementation of practices for mutual benefit to play a role in the long-term sustainability of the business. – Commitments are established with each of the stakeholders, generating a value promise in a win-win setting. Commitments include stakeholders' perceptions, needs and expectations, as well as the company's needs, criteria and strategic aspects. Alignment is achieved by associating the commitments with corporate objectives, making sure they are a part of the business strategy.	– Environmental strategy to go beyond what is required by law and become a world-class Company, recognised, among other things, for the commitment to environmental care. – Eco-efficiency: It aims to progressively reduce the environmental impacts of the operation and the investment projects. Specifically, in the reduction of discharges and solid waste. – As a part of biodiversity program, in 2011, Ecopetrol has undertaken the implementation of a project to evaluate populations of Crocodylus Acutus
Petrobras Brasilia	– Petrobras turns people and their development into a performance differential of Petrobras. – Biofuel production also allows for jobs and income generation on the farmland, taking advantage of Brazil's favourable conditions, such as its climate, water and a major agricultural frontier to be explored, without any need to advance into forest areas or demarked reserves. – encourages its suppliers to also adopt environmentally responsible attitudes and practices through requirements imposed in the supplier corporate register	– Biofuel production: alternative energies led us to develop biodiesel and the HBio process, and enhanced the ethanol trade. – To reinforce the performance in this segment, the company created Petrobras Biocombustível. – Developed proprietary technology for biodiesel production. – Renewable Fuel: Blending ethanol to gasoline also helped Brazil become one of the first countries to eliminate lead from gasoline.
OIL Ltd	– gives top priority towards the all-round development of the people residing in and around the Company's area of operation. – Makes an attempt to invest technology with a human face. – Been providing financial assistance to all-round development of education from primary to higher education.	– Usage of information and technology services in the operational areas.

contd...

...contd...

	– Company has provided financial assistance towards construction of school building, libraries, auditoriums, teachers' common room, etc., in over 350 institutions in the company operational areas in upper Assam.	
ONGC Ltd	– To provide work bench to young, talented, innovative, restless minds for proving and actualising worthy ideas.	– Harness science and technology to meet national energy needs of tomorrow in a clean and sustainable manner. – To develop clean, practical and affordable energy solutions through research & development as well as demonstration leading to commercialisation

Conclusions

The aim of this study is to provide an overview framework for corporate governance models, CSR and sustainability practices. The observed behaviour of Chinese oil companies is the result of a complex interplay between individuals and groups associated with the firms, and whose interests are not always aligned, and where commercial incentive is the main driver. It is a case (Sinopec, for instance) that shows governance system is in fact driven by the intensity of the products dealt in terms of national consumption and percentage of spend on GDP (gross domestic product). Although they are owned by government, they are not government run.

In case of LUKOIL, it is interesting to know that the 11-member Board is elected by the General Shareholders Meeting by cumulative voting (the candidates for whom the largest number of votes was cast are deemed elected) by including the election of the board members in the agenda of the annual General Shareholders Meeting. Fees and compensations to all board members are established by the General Shareholders Meeting. The chairperson is elected by the Board, and based on the recommendation of the board whereas chairperson appoints the secretary of the Board and elects committee members.

To make the investments more appealing, dividend policy is based on the principle that dividend payments should be no less than 15 per cent of net profit determined on the basis of the company's consolidated financial statements prepared in accordance with US Generally Accepted Accounting Principles (GAAP). The intention is to ensure consistent growth of profit and/or growth of the share of dividend payments in net profit. The LUKOIL Group's strategic objective is to ensure dynamic sustainable development corresponding to that of global oil majors in terms of their efficiency and competitiveness. LUKOIL Group social responsibilities consist in an adequate working environment and appropriate remuneration, environmental safety and preservation of the cultural heritage.

In case of Petrobras Brasilia, the Board of Directors features a collective and autonomous body within its prerogatives and responsibilities, as provided by law and the Articles of Incorporation. It consists of 10 directors who are elected at the General Ordinary Meeting for a one-year term, with re-election allowed. Seven of whom are appointed by the controlling shareholder, one by minority common shareholders, one by preferred shareholders and one by employees. The Executive Board of Directors undertakes the company's business, pursuant to the mission, goals, strategies and guidelines established by the board of directors. It comprises a chairperson and seven directors elected by the Board of Directors, with three-year terms, re-election being permitted, and may be dismissed at any time. Among the members of the Executive Board, only the president is a member of the Board of Directors without, however, presiding over the body.

Unlike, LUKOIL and Petrobras, the boards in Indian companies are constituted by the ministry based on the selection made by the public enterprise selection board which are finally approved by the cabinet committee on appointments. The recommendations made by the public enterprise selection board (PESB) have to be endorsed by the concerned administrative ministry before being considered by the cabinet committee on appointments. This system of selection procedure is cumbersome and deleterious for the healthy functioning of an enterprise. The reasons for this being so are many: one, the PESB normally forwards a panel of two-three incumbents for the appointment. The ministry should be endorsing the name of the candidate who is first in order but the ministry could even recommend the second or third candidate. The ministry may even disagree with the panel and may ask fresh selections. The cabinet committee on appointment takes its own time to make the final decision of recommending the candidate to the cabinet. Before considering the name, vigilance report on the conduct of the candidate has to be furnished to the cabinet committee on appointment. If all this goes smoothly, it takes about 8-12 months before the appointment takes place. This is indicative of corporate governance issues in India and are in line with the issues pointed by Scrimgeour and Duppati (2014).

The case of the five companies demonstrates seriousness in their approach towards CSR and sustainability practices through their strategy and implementation. Inspite of differences in the geographical location and board structures, the five cases show similarities in their behaviour towards social elements, sustainability practices including energy-efficiency measures, developing alternate energy sources like biofuel and renewable energy. The expectations and demands of the society towards firms in taking up responsibility for all stakeholders involved is increasing and the study, therefore, provides understanding about the approaches of the five corporate towards environment and society.

References

Becht, M. and A. Roell (1999). "Blockholdings in Europe: An International Comparison", *European Economic Review* 43(4–6): 1049-56.

Hall, P.A. and D. Soskice (2001). "An Introduction to Varieties of Capitalism", in P.A. Hall and D. Soskice (eds.), *Varieties of Capitalism: The Institutional Foundations of Comparative Advantage*. Oxford: Oxford University Press pp.1-70.

La Porta, R., F. Lopez-de-Silanes, A. Shleifer, and R.W. Vishny (1998). "Law and Finance", *Journal of Political Economy* 106(6): 1113-55.

Scrimgeour, F. and Geeta Duppati (2014). "Corporate Governance in the Public Sector: Dimensions; Guidelines and Practice in India and New Zealand", *Corporate Ownership and Control* 11(2): 364-77.

Shleifer, A. and R.W. Vishny (1997). "A Survey of Corporate Governance", *Journal of Finance* 52(2): 737-83.